# Analytical Solution Methods for Boundary Value Problems

# Analytical Solution Methods for Boundary Value Problems

**ANATOLY S. YAKIMOV**
Department of Physical and Computing Mechanics of National Research, Tomsk State University, Russia

ELSEVIER

AMSTERDAM • BOSTON • HEIDELBERG • LONDON
NEW YORK • OXFORD • PARIS • SAN DIEGO
SAN FRANCISCO • SINGAPORE • SYDNEY • TOKYO
Academic Press is an imprint of Elsevier

AP

Academic Press is an imprint of Elsevier
125 London Wall, London EC2Y 5AS, UK
525 B Street, Suite 1800, San Diego, CA 92101-4495, USA
50 Hampshire Street, 5th Floor, Cambridge, MA 02139, USA
The Boulevard, Langford Lane, Kidlington, Oxford OX5 1GB, UK



**Library of Congress Cataloging-in-Publication Data**
A catalog record for this book is available from the Library of Congress

**British Library Cataloguing in Publication Data**
A catalogue record for this book is available from the British Library

ISBN: 978-0-12-804289-2

For information on all Academic Press publications
visit our website at https://www.elsevier.com/



*Publisher:* Nikki Levy
*Acquisition Editor:* Graham Nisbet
*Editorial Project Manager:* Susan Ikeda
*Production Project Manager:* Poulouse Joseph
*Designer:* Greg Harris

# CONTENTS

# ABOUT THE AUTHOR

**Anatoly Stepanovich Yakimov** is a Doctor of Sciences, a senior research scientist and Professor of the Department of Physical and Computing Mechanics of National Research, Tomsk State University, Russia.

He graduated from the Mechanical-Mathematical Faculty of Tomsk State University in 1970. In 1981, he defended his candidate dissertation in physical and mathematical sciences on a specialist subject, in the Scientific-Research Institute of Applied Mathematics and Mechanics at Tomsk State University (speciality area 01.02.05 was mechanics of liquid, gas, and plasma). In 1999, he defended his Doctor thesis on the topic "Mathematical Modeling and the Numerical Solution of Some Problems of Heat and Mass Transfer and Thermal Protection" (speciality area 05.13.16 is application of computer facilities, mathematical modeling, and mathematical methods in scientific researches and 01.02.05) in Tomsk State University.

He is the author of one joint textbook, a monograph, and 70 scientific publications (not including abstracts) devoted to mathematical modeling of some problems of the thermal protection and working out of mathematical technology for solution of equations of mathematical physics. The last one is reflected in the following textbooks:

1. Grishin AM, Zinchenko VI, Efimov KN, Subbotin AN, Yakimov AS. The iterative-interpolation method and its appendices. Tomsk: Publishing House of Tomsk State University; 2004. 320 p.
2. Yakimov AS. Analytical method of solution of boundary problems. Tomsk: Publishing House of Tomsk State University; 2005. 108 p.
3. Grishin AM, Golovanov AN, Zinchenko VI, Efimov KN, Yakimov AS. Mathematical and physical modeling of thermal protection. Tomsk: Publishing House of Tomsk State University; 2011. 358 p.
4. Yakimov AS. The analytical method solution mathematical physics some equations. Tomsk: Publishing House of Tomsk State University; 2007. 150 p.
5. Yakimov AS. Analytical method of the solution of boundary problems. 2nd ed. Tomsk: Publishing House of Tomsk State University; 2011. 199 p.

The development of mathematical technology of solution of equations of mathematical physics is reflected in about one-quarter (17) of all publications, nine of which are devoted to analytical method of solution of boundary problems.

Based on the method of quasi-linearization, operational calculation and splitting onto the spatial variables, exact and approximate analytical solutions of the equations in private derivatives of the first and second order are obtained. Conditions of the unequivocal resolvability of nonlinear boundary problems are found and an estimation of speed of convergence of the iterative process is determined.

All exact and approximate formulas in solutions of equations of mathematical physics (that are considered in the present book) are obtained in an explicit form without the use of the theory of rows.

# INTRODUCTION

Mathematical modeling of processes in various areas of science and technology in many cases represents the unique way of reception of new knowledge and new approaches to technological solutions.

During the last decades of the 20th century, considerable progress was achieved in the solutions to many problems (in space, atomic engineering, biology, etc.) thanks to the application of computing algorithms and the COMPUTER.

The great number of problems in physics and techniques leads to linear and nonlinear boundary problems (the equations of mathematical physics). According to scientists' estimations, the effect received by the perfection of a solution algorithm can amount to a 40% or greater increase in productivity of the COMPUTER. But the signal possesses a maximum speed of distribution, ie, the speed of light. Therefore, the growth of speed of uniprocessor COMPUTERS is limited. At the same time, the effect of an increase toward perfection of an algorithm, theoretically, is unlimited.

In some cases in solutions of equations of mathematical physics, the analytical or the approximate analytical methods can compete with the numerical methods. This concerns not only the simplified mathematical statements of boundary problems (constant transfer factors, absence of nonlinear sources, one-dimensionality on a spatial variable, etc.), but also the mathematical models describing real physical processes (nonlinear, in three-dimensional space, etc.).

In the mathematical modeling of problems of mechanics: heat- and mass-transfer [1, 2], unsteady thermal streams in cars, isolation [3]; electronics: calculation of the electric contours [4], etc.—there are challenges in the formulation of solution in equations of mathematical physics. In the solution of boundary problems for the quasi-linear equation of heat conductivity [5] analytical formulas in some special cases are received (infinite range of definition on spatial coordinate or absence source). Exact analytical solutions on a final interval in the space are received only for one-dimensional linear transfer equation with a source [3, 4, 6]. However, in practice the solution of the nonlinear boundary problems [1, 2, 7], by application of basically numerical methods is more interesting.

The problem of acceptance of the analytical solutions of boundary problems for a nonlinear equation of heat conductivity was considered in [5] and it was noticed that to find the analytical solution of the equation heat conductivity on a final point with any source is impossible. This result is especially true in the spatial case. Methods of solution of nonlinear one-dimensional boundary problems for sources of a special kind are presented in a review [5].

For the solution of one-dimensional nonlinear ordinary differential equations of the second order [7] the method of quasi-linearization is offered. Using this method, the solution of the nonlinear problem is reduced to the sequence of linear problems solution that represents the development of the well-known Newton's method and its generalized variant offered by Kantorovich [8]. Otherwise, the method of quasi-linearization is an application to a nonlinear function generated by a nonlinear boundary problem, the method of Newton-Kantorovich.

In the book, on the basis the method of quasi-linearization and Laplace integral transformation [6], the analytical solution of the first boundary problem for the nonlinear one-dimensional equation of parabolic type [2] in the final point with a source has been obtained. The work [9] offers an iterative-interpolation solution method of one-dimensional linear and nonlinear boundary problems. We mention also the article [10], in which Newton-Kantorovich's method together with a method of grids was applied to the solution of one-dimensional boundary problems. Then in [11] convergence of sequence in fundamental solutions of equations of heat conductivity is proved and also examples of solutions of some of equations of mathematical physics are given. It is necessary to mention that the idea of the method of quasi-linearization is very close to the idea of methods [9–11]. In all cases, the method of consecutive approaches is used. The difference is that finally, in works [10, 11], various final differences are used, and in the present book the linear problem is solved analytically and thus convergence of the iterative process remains square-law.

In the numerical solution of multidimensional problems of mathematical physics, the splitting methods [12–14] turned out to be effective. Particularly the locally one-dimensional scheme of splitting [12, 13] is offered for the solution of the multidimensional equation of heat conductivity in a combination with analytical (constant factors) and numerical methods.

The purpose of the present book is to develop mathematical technology solutions of boundary problems on the basis of the method of quasi-linearization, operational calculation and the locally one-dimensional scheme of splitting on to spatial coordinates (in the three-dimensional case), to receive conditions of unequivocal resolvability of nonlinear boundary problems and to find an estimation of the speed of convergence of the iterative process.

In Chapter 1 of the book on the basis of operational calculation, the exact formula is developed while solving three-dimensional equations in private derivatives of the first order and by means of operational calculation and the locally one-dimensional scheme of splitting the analytical solution is found for three-dimensional elliptic equations with constant factors. On the basis of trial functions the result of comparison with the known numerical method is given.

In Chapter 2, first of all on the basis the method of quasi-linearization and Laplace integral transformation, the approach of the analytical solution of nonlinear boundary problem for one-dimensional transfer equation is established. Then, using the method of quasi-linearization, the locally one-dimensional scheme of splitting and operational calculation, analytical formulas are found in the solution of three-dimensional nonlinear transfer equations. Existence, uniqueness of sequence of approximates to the required solution of the boundary problem is given and the estimation of the speed of convergence of the iterative process is also considered. Results of test checks are stated and calculation comparison on given mathematical technology with the numerical solution of a problem is performed.

In Chapter 3 of the book, by means of the method of quasi-linearization and Laplace integral transformation, the approximate analytical solution of the first boundary problem for one-dimensional equation of parabolic type has been presented and the approximate analytical formulas in the solution of nonlinear boundary problems in the spatial case [15, 16] are given. Following the algorithm of Chapter 2, the estimation of speed of convergence of the iterative process in the solution of the first boundary problem has been derived. By test examples accuracy offered by mathematical technology is estimated and the result of a comparison with a well-known numerical method is given.

In Chapter 4, the algorithm of the solution of the conjugate problem of heat exchange [17] is determined and the approach of the analytical formulas for nonlinear boundary problems in one-dimensional and spatial statements are presented. For the nonlinear boundary problem, the estimation of speed of convergence of the iterative process is found in the conjugate statement.

In mathematical modeling, heat transfer in high-intensity processes [17], is determined by electromagnetic oscillations [18] problem solutions of the telegraph equation.

In the last chapter we are presented with approximate analytic formulas for solving partial differential equations of the second-order hyperbolic type. For the nonlinear one and three-dimensional boundary value problem an estimate of speed of convergence of the iterative process is obtained. Utilizing the method of test functions results of the test checks in mathematical modeling and a comparison with the known numerical method and the analytical solution are also presented.

Comparison of the accuracy of the solution of the one-dimensional problem of heat exchange under the offered mathematical technology with the known analytical method of this problem has resulted. In the spatial case, the method of trial functions receives results in test checks of mathematical technology and comparison with a known numerical method and the analytical solution is given.

All cores of the exact and approached analytical formulas at the solution of equations of mathematical physics (with sources, variable by factors) are received in an explicit form without the use of mathematical technology [18], connected with concept of a row [19] that is important.

The represented mathematical technology solution of boundary problems does not extend to unlimited ranges of definition in space.

In summary, we acknowledge that the book is written on the basis of the results of research work developed in the laboratory, on "Modelling and the Forecast of Accidents," scientifically researched in Tomsk State University and by the Chair of the Physical and Computing Mechanics of the mechanics-mathematical faculty at Tomsk State University. The material presented to the reader was used by the author in a special course "Numerical Methods Mechanics of the Continuous Environment" between 2005 and 2009.

## CHAPTER 1

# Exact Solutions of Some Linear Boundary Problems

The purpose of this chapter is to work out technology for the solution of the three-dimensional linear transfer equation on the basis of operational calculation and to obtain analytical formulas for the solution of a three-dimensional elliptic equation with constant coefficients.

For early publications on operational calculation, for example, on two variables, refer to the 1930s and 1950s (R. Ember, G. Dech, V.A. Ditkin, etc.). The operational calculation theory, based on the application of non-one-dimensional Laplace integral transformation, follows from the general theory as a special case in consideration of operators transformed by Laplace.

Later, Dech [4] recommended, for the solution of the multidimensional linear transfer equation, to apply the Laplace integral transformation as much as the dimension of this equation is. Solving the last one by Laplace integral transformation and consistently finding originals by known tables [4, 6], we definitively obtain the problem solution.

In pursuing this analysis we consider that for the boundary problems there are only continuous solutions with the necessary number of continuous derivatives of space and time.

The analytical solution is suggested in Section 1.2 by means of mathematical technology from [16] the solution of three-dimensional elliptic equations with constants coefficients. At the end of each point on the basis of trial functions the results are compared with those obtained with a known numerical method.

## 1.1 ANALYTICAL METHOD OF SOLUTION OF THREE-DIMENSIONAL LINEAR TRANSFER EQUATIONS

### Statement of a Problem and a Method Algorithm

Let us attempt to find the solution of linear transfer equations [1, 20]:

*Analytical Solution Methods for Boundary Value Problems*
http://dx.doi.org/10.1016/B978-0-12-804289-2.00001-6

$$\frac{\partial u}{\partial t} + \sum_{m=1}^{3} c_m \frac{\partial u}{\partial x_m} = f(t, x), \quad u \geq 0 \tag{1.1}$$

with initial value

$$u|_{t=0} = w(x). \tag{1.2}$$

Let the definiteness signs on sizes $c_m$ = const be known in advance, for example: $c_m > 0$, $m = 1, 2, 3$ at $0 \leq t < t_k$ $(0 < t_k < \infty)$. Then boundary conditions in a parallelepiped $R$ : $[x = (x_1, x_2, x_3),\ 0 \leq x_m < S_m\ (0 < S_m < \infty,\ m = 1, 2, 3)]$ are set as [1, 20]

$$u|_{x_1=0} = g_1(t, x_2, x_3), \quad u|_{x_2=0} = g_2(t, x_1, x_3),$$
$$u|_{x_3=0} = g_3(t, x_1, x_2). \tag{1.3}$$

Let's assume everywhere:

1. The problem (1.1)–(1.3) has an unambiguous solution $u(x, t)$, which is continuously in the closed region $\overline{R_t} = \overline{R} \times [0 \leq t \leq t_k]$, $\overline{R} = R + \Gamma, \Gamma$—the boundary surface also has continuous derivatives $\partial u/\partial t$, $\partial u/\partial x_j$, $j = 1, 2, 3$.
2. The following conditions are provided: $g_m \geq 0, m = 1, 2, 3, w \geq 0$, $f(x, t) \geq 0$ and only the positive solution of a boundary problem (1.1)–(1.3) can be found.
3. Functions $w$, $g_m$, $m = 1, 2, 3, f(x, t)$ are continuous in the region under consideration.

Let's consider Laplace integral transformation [4, 6, p. 314] and extend it to the multidimensional case by analogy with the two-dimensional in the form of

$$U(p, x) = \int_0^\infty \exp(-pt) u(t, x_1, x_2, x_3)\, dt,$$

$$U_1(p, q, x_2, x_3) = \int_0^\infty \exp(-qx_1) U(p, x)\, dx_1,$$

$$U_2(p, q, s, x_3) = \int_0^\infty \exp(-sx_2) U_1(p, q, x_2, x_3)\, dx_2,$$

$$U_3(p, q, s, r) = \int_0^\infty \exp(-rx_3) U_2(p, q, s, x_3)\, dx_3, \tag{1.4}$$

where $p$, $q$, $s$, $r$—are complex parameters, and indexes 1, 2, 3 at $U_1$, $U_2$, $U_3$ refer to Laplace integral transformation on spatial variables. Further functions for which Laplace integral transformation absolutely converges [6] are considered. The real parts of numbers $p$, $q$, $s$, $r$ are considered as positive, that is $\mathrm{Re}\, p > 0$, $\mathrm{Re}\, q > 0$, $\mathrm{Re}\, s > 0$, $\mathrm{Re}\, r > 0$. We assume that the basic properties of multidimensional Laplace integral transformation are generalized and are similar to corresponding properties of two-dimensional Laplace transformation [6].

So the Laplace integral transformation to Eq. (1.4) gives

$$L_r^{-1}[U_3] = U_2(p,q,s,x_3), \quad L_s^{-1}[U_2] = U_1(p,q,x_2,x_3),$$
$$L_q^{-1}[U_1] = U(p,x), \quad L_p^{-1}[U] = u(t,x). \tag{1.5}$$

Let's apply Laplace integral transformation (1.4) to the Eq. (1.1). We assume that the required solution $u(t,x)$, and also its derivatives entering into the Eq. (1.1), satisfy Laplace integral transformation conditions on $t$ and on $x$, and its growth degree on $t$ functions $u(t,x)$ and its derivatives do not depend on $x$. Similarly, the growth degree on $x_1$ functions $u(t,x)$ and its derivatives do not depend on $t$, $x_2$, $x_3$, etc. Subsequently, multiplying both sides of the Eq. (1.1) by $\exp(-pt)$ and integrating on $t$ from 0 to $\infty$, and then multiplying on $\exp(-qx_1)$ and integrating on $x_1$ from 0 to $\infty$, etc., we obtain (using an integration rule by parts)

$$c_1\frac{\partial U}{\partial x_1} + c_2\frac{\partial U}{\partial x_2} + c_3\frac{\partial U}{\partial x_3} + pU(p,x) = u(0,x) + F(p,x);$$
$$c_1 q U_1(p,q,x_2,x_3) - c_1 U(p,0,x_2,x_3) + c_2\frac{\partial U_1}{\partial x_2} + c_3\frac{\partial U_1}{\partial x_3}$$
$$+ pU_1(p,q,x_2,x_3) = U(0,q,x_2,x_3) + F_1(p,q,x_2,x_3);$$
$$c_2 s U_2(p,q,s,x_3) - c_2 U_1(p,q,0,x_3) + c_1 q U_2(p,q,s,x_3) - c_1 U_1(p,0,s,x_3)$$
$$+ c_3\frac{\partial U_2}{\partial x_3} + pU_2(p,q,s,x_3) = U_1(0,q,s,x_3) + F_2(p,q,s,x_3);$$
$$c_3 r U_3(p,q,s,r) - c_3 U_2(p,q,s,0) + c_2 s U_3(p,q,s,r) - c_2 U_2(p,q,0,r)$$
$$+ c_1 q U_3(p,q,s,r) - c_1 U_2(p,0,s,r) + pU_3(p,q,s,r) = U_2(0,q,s,r)$$
$$+ F_3(p,q,s,r). \tag{1.6}$$

Let's transform the Eq. (1.6), collecting similar summands at $U_3$. As a result we obtain

$$U_3(p,q,s,r) = [U_2(0,q,s,r) + c_3 U_2(p,q,s,0) + c_2 U_2(p,q,0,r)$$
$$+ c_1 U_2(p,0,s,r) + F_3(p,q,s,r)]/(p+a),$$
$$a = c_1 q + c_2 s + c_3 r. \tag{1.7}$$

To find originals in Eq. (1.7) we use the table [4] and the formula from [6, p. 151]:

$$\eta(t)\exp(-p\tau) = \begin{cases} 0, & t < \tau, \quad \tau \geq 0, \\ \eta(t-\tau), & t \geq \tau; \end{cases} \tag{1.8}$$

$$L_p^{-1}[\exp(-p\tau)U(p)] = u(t-\tau), \quad \tau > 0,$$

$$L_p^{-1}\left[\frac{U(p)}{p+a}\right] = \int_0^t \exp(-a\tau)u(t-\tau)\,d\tau. \tag{1.9}$$

Using Laplace integral transformation (1.5) consecutively, for example, to the second summand of the right-hand side of the Eq. (1.7) and using Eq. (1.9), we find

$$\begin{aligned} &L_p^{-1}\left[\frac{U_2(p,q,s,0)}{p+a}\right] = \int_0^t \exp(-a\tau)U_1(t-\tau,q,s,0)\,d\tau = V_1, \\ &L_q^{-1}[V_1] = \int_0^t \exp[-\tau(c_2 s + c_3 r)]U(t-\tau, x_1-\tau c_1, s, 0)\,d\tau = V_2, \\ &L_s^{-1}[V_2] = \int_0^t \exp(-\tau c_3 r)g_3(t-\tau, x_1-\tau c_1, x_2-\tau c_2, 0)\,d\tau = V_3, \\ &L_r^{-1}[V_3] = I_3, \end{aligned} \tag{1.10}$$

where

$$I_3 = \int_0^t g_3(t-\tau, x_1-\tau c_1, x_2-\tau c_2, 0-\tau c_3)\,d\tau, \quad \tau c_3 \leq 0,$$

$$I_3 = 0, \quad \tau c_3 > 0.$$

Let's note that the order of restoration of originals in Eqs. (1.7) and (1.10) on parameters $p, q, s, r$ is set by operators (1.5). To receive the last expression ($I_3$) in a chain (1.10), functional dependence is used (1.8).

The originals (1.5), (1.8), and (1.9) for other summands in the right-hand side (1.7) are restored in the same way. As a result the required solution of the Eq. (1.1) may be written as:

$$\begin{aligned} u(t,x) = w(0, x_1 - tc_1, x_2 - tc_2, x_3 - tc_3) + \int_0^t f(t-\tau, \\ x_1 - \tau c_1, x_2 - \tau c_2, x_3 - \tau c_3)\,d\tau + \sum_{m=1}^{3} I_m c_m, \end{aligned} \tag{1.11}$$

where

$$I_1 = \int_0^t g_1(t-\tau, 0-\tau c_1, x_2-\tau c_2, x_3-\tau c_3)\, d\tau, \quad \tau c_1 \le 0,$$

$$I_1 = 0, \quad \tau c_1 > 0;$$

$$I_2 = \int_0^t g_2(t-\tau, x_1-\tau c_1, 0-\tau c_2, x_3-\tau c_3)\, d\tau, \quad \tau c_2 \le 0,$$

$$I_2 = 0, \quad \tau c_2 > 0.$$

It is obvious that calculation by formulas (1.11) is possible, if $w$, $g_m$, $m = 1, 2, 3$ from Eqs. (1.2) and (1.3) and $f$ from Eq. (1.1) are set by analytical expressions from spatial coordinates and time.

While finding the integrals entering into the Eqs. (1.10) and (1.11), we use Simpson's formula [21] which has the fourth order of accuracy. It allows split-hair accuracy if the fourth derivative subintegral function is small. As the last one is determined analytically (unlike final differences, where the required function is unknown in advance), it is possible to estimate it. At a small number of nodes (~9) it is possible to use, for example, Newton's-Kotesa formula [21] which has the tenth order of accuracy.

## Example of Test Calculation

In the one-dimensional case, for example for $c_1 = 1$ and trial function $u = (1+t)(1+x_1)$, the source in the Eq. (1.12)

$$\frac{\partial u}{\partial t} + \frac{\partial u}{\partial x_1} = f(t, x_1), \quad x_1 \in (0,1),\ t > 0 \tag{1.12}$$

looks like $f = 2 + t + x_1$. Then, knowing $u|_{t=0} = 1 + x_1, u|_{x_1=0} = 1 + t$ and

$$w(0, x_1 - t) = 1 + x_1 - t, \quad \int_0^t f(t-\tau, x_1-\tau)\, d\tau$$

$$= \int_0^t (2 + t + x_1 - 2\tau)\, d\tau,$$

$$I_1 = \int_0^t g_1(t-\tau, 0-\tau)\, d\tau = \int_0^\tau g_1(\tau-\tau, 0-\tau)\, d\tau = \tau g_1(0) = \text{const}$$

($I_1 = 0$ at $\tau > 0, I_1 \ne 0$ at $\tau \le 0, t = \tau$), the solution of a problem of type (1.1)–(1.3) in the one-dimensional statement is written as ($\tau g_1(0) = 0$ at $\tau \le 0$, as $\tau$ from physical reasons is a positive value):

$$u(t, x_1) = 1 + x_1 - t + \int_0^t (2 + t + x_1 - 2\tau)\, d\tau + \tau g_1(0). \tag{1.13}$$

Let's substitute $u(t, x_1)$ in the Eq. (1.12); then we have for $g_1(0) = 1$

$$\frac{\partial u}{\partial t} = -1 + 2 + 2t + x_1 - 2t + 0 = 1 + x_1.$$

For integrals $J_1 = \int_0^t z(\tau)\, d\tau$, $J_2 = \int_{v(t)}^{w(t)} y(t,\tau)\, d\tau$ in Eq. (1.13) known rules [19] are used:

$$\frac{\partial J_1}{\partial t} = z(t), \quad J_2' = \int_{v(t)}^{w(t)} y_t'(\tau, t)\, d\tau + w'(t)y[w(t), t] - v'(t)y[v(t), t],$$

$$\frac{\partial u}{\partial x_1} = 1 + \int_0^t \frac{\partial}{\partial x_1}(2 + t + x_1 - 2\tau)\, d\tau + 0 = 1 + t.$$

Here, functions $f(t, x_1)$ and $\partial f / \partial x_1$ are obviously continuous in a rectangle $R : (0 \le t \le t_k, 0 \le x_1 \le S_1)$. As a result we obtain the identity

$$\frac{\partial u}{\partial t} + \frac{\partial u}{\partial x_1} = 2 + t + x_1 \equiv f(t, x_1).$$

The solution (1.11) for a three-dimensional transfer Eq. (1.1) is checked in a similar way. However, it is not always possible to integrate a source.

Accuracy of the obtained analytical expression (1.11) and correctness of algorithm calculation in a spatial case we can establish while solving a three-dimensional differential problem (1.14) from [22] with boundary conditions (1.15)

$$\frac{\partial u}{\partial t} + \sum_{m=1}^{3} c_m \frac{\partial u}{\partial x_m} = f(t, x), \quad u \ge 0, \quad x_m \in (0, 1),\ t > 0, \tag{1.14}$$

$$h(x) = 1 + x_1^z + x_2^z + x_3^z, \quad f = 4h(x)t^3 + z(1 + t^4) \sum_{m=1}^{3} c_m x_m^{z-1};$$

$$u|_{x_1=0} = (1 + t^4)(1 + x_2^z + x_3^z), \quad u|_{x_2=0} = (1 + t^4)(1 + x_1^z + x_3^z),$$
$$u|_{x_3=0} = (1 + t^4)(1 + x_1^z + x_2^z), \quad u|_{t=0} = h(x). \tag{1.15}$$

The exact obvious solution of the regional problem (1.14) and (1.15) is known in advance: $u = h(x)\ (1 + t^4)$ at $0 \le t \le t_k$. First of all, the problem solution is considered (1.14) and (1.15) in a one-dimensional statement. In this case it is possible to compare the accuracy of the derived formulas (1.11) (further to approach 1) with the known analytical solution (1.16) (further to approach 2), found in [6] operational calculation on a basis of the two-dimensional Laplace integral transformation:

**Table 1.1** Dependence of the maximum relative error on time

| $t$ | 0.1 | 0.5 | 1 | 2 | 5 |
|---|---|---|---|---|---|
| $\varepsilon_1$, % | 0.067 | 0.33 | 0.66 | 1.33 | 3.33 |
| $\varepsilon_2$, % | 0.02 | 0.51 | 1.567 | 1.23 | 0.75 |

$$u(t,x_1)=\begin{cases} w(x_1-c_1t)+\int_0^t f(t-\tau,x_1-c_1\tau)\,d\tau, & y_1>t,\\ g(t-y_1)+c_1^{-1}\int_0^{x_1} f(t-\xi/c_1,x_1-\xi)\,d\xi, & t>y_1, \end{cases} \tag{1.16}$$

where $y_1 = x_1/c_1$.

The following values of input data are taken: $S_1 = 1$, $c_1 = 1$, $M = 51$, $z = 4$, $N_1 = 11$, $\Delta t = t/(M-1)$: $N_1$ is a number of calculation nodes in space, $\Delta t$ is the time step in calculation of integrals in the Eqs. (1.11) and (1.16) with the Simpson formula. It is obvious that accuracy in the calculation with formulas (1.11) and (1.16) depends on the accuracy of calculation of integral with quadrature formula. The program is made on Fortran-90, the calculation was made on COMPUTER Pentium 3 (130 MHz, compiler PS 4), with double accuracy.

In Table 1.1 the maximum relative time error is given as $\varepsilon = |u - \bar{u}|$ $100\%/u$ ($u$ is exact, $\bar{u}$ is the analytical solution); $\varepsilon_1$ corresponds to the result of the one-dimensional ($m = 1$) boundary problem solution (1.14) and (1.15) approach 1, and $\varepsilon_2$ to approach 2.

Table 1.1 shows the accuracy of calculation with both approaches practically coinciding. It is necessary to mention that using approach 1 it is possible to improve the accuracy of a problem solution, reducing $\Delta t$ (increasing $M$). At the same time, in approach 2, to receive good accuracy, it is necessary to reduce a step on a spatial variable, as the second integral in Eq. (1.16) where a variable top limit is on $x$. Also, in this case for even nodes on space it is necessary to use the trapezoid rule for its calculation.

Now we will compare approach 1 in a three-dimensional case with numerical calculation of a problem (1.14) and (1.15) on technology from [22] at $S_m = 1$, $c_m = 1$, $N_m = 11$, $m = 1, 2, 3$, $z = 4$. In order to achieve almost equal accuracy, for example, during the final moment of time $t_k = 5$ in a method [22] approach 3 was necessary $\Delta t = 0.005$. In Table 1.2 $\varepsilon_3$ corresponds to the result of a problem solution with approach 3, and $\varepsilon_1$ with approach 1 at $\Delta t = 0.025$ ($M = 201$).

Thus, the time calculation ($t_p$) for approach 1 amounted to $t_p = 21$ s, whereas for approach 3, it was $t_p = 5.5$ min (15 times longer).

**Table 1.2** Dependence of the maximum relative error on time

| $t$ | 0.1 | 0.5 | 1 | 2 | 5 |
|---|---|---|---|---|---|
| $\varepsilon_1$, % | 0.05 | 0.25 | 0.5 | 1.0 | 2.5 |
| $\varepsilon_3$, % | 0.44 | 1.38 | 5.66 | 6.04 | 2.6 |

## 1.2 THE EXACT SOLUTION OF THE FIRST BOUNDARY PROBLEM FOR THREE-DIMENSIONAL ELLIPTIC EQUATIONS

We know [11, 23–25] the iterative methods of solution of the multidimensional Poisson's equation. Some methods use a parameter of relaxation $\omega(0 < \omega < 2)$, which is necessary to select in solutions of elliptic equations (1.17). Other iterative schemes can be treated [11, 24] as methods of establishments (at $t \to \infty$) for corresponding a non-stationary equation with stationary (not dependent from time) boundary data. In this case, it is necessary to find the optimum iterative parameter $\tau_0$ from a condition in which difference solutions will be stationary one for the least number of steps. The problem of finding $\omega$ and $\tau_0$ for equations of a type (1.17) is quite difficult and not always possible to solve, even for constant transfer coefficients.

In numerical solutions of multidimensional equations of mathematical physics, the splitting methods [11–14] turned out to be effective. In particular, the locally-one-dimensional scheme of splitting [13] is offered for solutions of multidimensional equation of heat conductivity in a combination with analytical (constant coefficients) and numerical methods.

Therefore association of operational calculation [6] is of particular interest (especially with a Laplace integral transformation) [12] for the solution of Eq. (1.17). Application of a Laplace integral transformation is connected with that unlike the general formula of solution of a non-uniform equation [26], the boundary conditions of the 2nd and the 3rd type are realized.

### Statement of a Problem and a Method Algorithm

Let us attempt to find the solution of a three-dimensional equation in private derivatives [2, 27]:

$$\sum_{m=1}^{3} \frac{\partial}{\partial x_m}\left(A_m \frac{\partial T}{\partial x_m}\right) + \sum_{m=1}^{3} B_m \frac{\partial T}{\partial x_m} + CT + f(x) = 0 \qquad (1.17)$$

in a parallelepiped $R : x = (x_1, x_2, x_3), (0 \le x_m < L_m, 0 < L_m < \infty,$ $m = 1, 2, 3)$ and for simplicity of the analysis in the absence of the mixed derivatives and at the boundary conditions of 1st type:

$$T|_{x_1=0} = g_1(0, x_2, x_3), \quad T|_{x_1=L_1} = q_1(L_1, x_2, x_3); \tag{1.18}$$

$$T|_{x_2=0} = g_2(x_1, 0, x_3), \quad T|_{x_2=L_2} = q_2(x_1, L_2, x_3); \tag{1.19}$$

$$T|_{x_3=0} = g_3(x_1, x_2, 0), \quad T|_{x_3=L_3} = q_3(x_1, x_2, L_3). \tag{1.20}$$

The equations of the type (1.17) are applied in the mechanics of inert and reacting environments [2]. Further we assume that $A_m = \text{const}, B_m = \text{const}, C = \text{const}, m = 1, 2, 3$. While solving, for example, the stationary equations of heat conductivity from the physical it is clear that heat conductivity coefficients in Eq. (1.17) are positive $(A_m > 0, m = 1, 2, 3)$. In a one-dimensional case (1.21) with constant coefficients and continuous right-hand side $f$ in a range of definition $R$ we can obtain a condition of the unequivocal resolvability in an explicit form. The corresponding Sturm-Liouville problem cannot be considered [28]. We will consider however that, the necessary condition of unequivocal resolvability of quasi-one-dimensional problems (1.21)–(1.23) is taken into account.

For a designation of stages of the intermediate problem solution (1.17) we will enter the top indexes: 1/3, 2/3, 1 [29] then, using the idea of splitting [12, 13], for the Eq. (1.17), we find

$$\begin{gathered} A_1 d^2 T^{(1/3)}/dx_1^2 + B_1 dT^{(1/3)}/dx_1 + f = 0, \\ T^{(1/3)}|_{x_1=0} = g_1, \quad T^{(1/3)}|_{x_1=L_1} = q_1; \end{gathered} \tag{1.21}$$

$$\begin{gathered} A_2 d^2 T^{(2/3)}/dx_2^2 + B_2 dT^{(2/3)}/dx_2 + \sigma_1 C T^{(1/3)} = 0, \\ T^{(2/3)}|_{x_2=0} = g_2, \quad T^{(2/3)}|_{x_2=L_2} = q_2; \end{gathered} \tag{1.22}$$

$$\begin{gathered} A_3 d^2 T^{(1)}/dx_3^2 + B_3 dT^{(1)}/dx_3 + \sigma_2 C T^{(2/3)} = 0, \\ T^{(1)}|_{x_3=0} = g_3, \quad T^{(1)}|_{x_3=L_3} = q_3, \end{gathered} \tag{1.23}$$

where $\sigma_m = 0.5, m = 1, 2, 0 < x_m < L_m, m = 2, 3$, for example, in the Eq. (1.21) are changed parametrically.

According to a splitting method [12, 13] the problem solution (1.17)–(1.20) is reduced to the consecutive solution of quasi-one-dimensional solutions (1.21)–(1.23). It is a question of the model convection-conduction head-conductivity. At the first stage, convection-conduction head-conductivity is switched off in directions of coordinates $x_2, x_3$, that is the problem is considered (1.21). Then we receive temperature distribution $T^{(1/3)}(x)$. Accepting it for intermediate, we switch off convection-conduction head-conductivity in directions of coordinates $x_1, x_3$, we consider the problem (1.22) and receive temperature distribution $T^{(2/3)}(x)$. Taking it again for intermediate, we switch off convection-conduction head-conductivity in directions of coordinates $x_1, x_2$ and consider the problem (1.23). Then finally we find the temperature $T^{(1)}(x)$, which coincides with the required value $T(x)$. According to this model process convection-conduction head-conductivity "is stretched" and occurs for three stages.

## Application of Integrated Laplace Transformation

We apply Laplace integral transformation (1.24) originally to the differential equation (1.21)

$$\overline{T}(p, x_2, x_3) = \int_0^\infty \exp(-px_1)T(x)\,dx_1 \quad (p = \tau + I\nu), \quad I = \sqrt{-1}, \tag{1.24}$$

excepting a derivative on $x_1$, replacing with its linear expression in relation to images of required function.

It is supposed that while calculating images on coordinates $x_j, j = 1, 2, 3$ we operate with functions, analytically continued on values $x_j > L_j$, by the law which they are defined in an interval $(0, L_j)$.

Further we consider functions for which Laplace integral transformation converges absolutely [6]. The valid part $p$ is considered positive, that is $\mathrm{Re}\, p > 0$. We assume that the required solution $T(x)$, and also it derivatives entering into the Eq. (1.21), satisfy existing conditions of Laplace integral transformation on $x$, and growth degree on $x_1$ functions $T(x)$ and its derivatives does not depend on $x_2, x_3$. In the same way growth degree on $x_2$ functions $T(x)$ and its derivatives does not depend on $x_1, x_3$, etc. Then using formulas from [6] and lowering an index for brevity's sake 1/3 above for $\overline{T}^{(1/3)}(p, x_2, x_3)$ and $T^{(1/3)}(x)$, we have for the image $\overline{T}$ and $(dT^{(1/3)}/dx_1)|_{x_1=0} = dg_1/dx_1$:

$$\begin{aligned} &A_1[p^2\overline{T}(p, x_2, x_3) - pg_1 - dg_1/dx_1] + B_1[p\overline{T}(p, x_2, x_3) - g_1] \\ &\quad + F(p, x_2, x_3) = 0. \end{aligned} \tag{1.25}$$

Let's transform the Eq. (1.25), collecting similar summands at $\overline{T}(p, x_2, x_3)$. As a result we find

$$\overline{T}(p, x_2, x_3) = A_1 p g_1/p(A_1 p + B_1) + (A_1 dg_1/dx_1 + B_1 g_1)/p(A_1 p + B_1) \\ - F(p, x_2, x_3)/p(A_1 p + B_1). \quad (1.26)$$

For application of Laplace integral transformation we present a denominator in the second summand of the right-hand side (1.26) in the form of

$$1/p(p + a_1) = [1/p - 1/(p + a_1)]/a_1, a_1 = B_1/A_1.$$

Then, using dependences from [6]: $L^{-1}[1/(p + a_1)] = \exp(-x_1 a_1)$, $L^{-1}[F(p)/p] = \int_0^{x_1} f(\xi)\, d\xi$, we restore the original for $T(x)$:

$$T(x) = g_1 + a_1^{-1} dg_1/dx_1 [1 - \exp(-x_1 a_1)] \\ - B_1^{-1} \int_0^{x_1} f(\xi, x_2, x_3)\{1 - \exp[-a_1(x_1 - \xi)]\}\, d\xi. \quad (1.27)$$

The derivative $dg_1/dx_1$ in expression (1.27) we will find using the second boundary condition from Eq. (1.21)

$$q_1 = T(L_1, x_2, x_3) = g_1 + a_1^{-1} dg_1/dx_1 [1 - \exp(-L_1 a_1)] \\ - B_1^{-1} \int_0^{L_1} f(\xi, x_2, x_3)\{1 - \exp[-a_1(L_1 - \xi)]\}\, d\xi.$$

Then expression for $T^{(1/3)}(x)$ of the first stage can be written as

$$T^{(1/3)}(x) = g_1 - B_1^{-1} \int_0^{x_1} f(\xi, x_2, x_3)\{1 - \exp[-a_1(x_1 - \xi)]\}\, d\xi \\ + [1 - \exp(-x_1 a_1)][1 - \exp(-L_1 a_1)]^{-1} \left\{ q_1 - g_1 \right. \\ \left. + B_1^{-1} \int_0^{L_1} f(\xi, x_2, x_3)(1 - \exp[-a_1(L_1 - \xi)])\, d\xi \right\}.$$

For simplicity of further calculations, we will enter Green's function $G$ [7, 9] and a designation

$$P_n = \frac{1 - \exp(-a_n x_n)}{1 - \exp(-a_n L_n)}, \quad n = 1, 2, 3;$$

$$G(x_1, y) = \begin{cases} P_1\{1 - \exp[a_1(y - L_1)]\} + \exp[a_1(y - x_1)] - 1, & 0 \le y \le x_1; \\ P_1\{1 - \exp[a_1(y - L_1)]\}, & x_1 \le y \le L_1. \end{cases}$$

As a result, the formula for $T^{(1/3)}(x)$ will be rewritten:

$$T^{(1/3)}(x) = g_1 + P_1(q_1 - g_1) + B_1^{-1} \int_0^{L_1} G(x_1, \xi) f(\xi, x_2, x_3)\, d\xi, \quad (1.28)$$

where $0 < x_m < L_m, m = 2, 3$ are changed parametrically.

Similarly Eqs. (1.25)–(1.28) it is possible to solve $T^{(2/3)}(x)$, $T^{(1)}(x)$ for problems (1.22) and (1.23):

$$T^{(n/3)}(x) = g_n + P_n(q_n - g_n) + b_n \int_0^{L_n} G(x_n, \xi) T^{(n-1)/3}(\xi)\, d\xi, \quad (1.29)$$

where $a_n = B_n/A_n, b_n = \sigma_{n-1}C/B_n, \sigma_{n-1} = 0.5, n = 2, 3, T^{(1/3)}(\xi) = T^{(1/3)}(x_1, \xi, x_3)$, $T^{(2/3)}(\xi) = T^{(2/3)}(x_1, x_2, \xi)$. If $C = 0$ (absence of "kinetic" source, a drain), we believe $b_2 = \sigma/B_2, b_3 = -\sigma/B_3, 0 < \sigma \leq 1$. At $n = 2, 3$ it is necessary in the formula for $G(x_1, \xi)$ to replace the bottom index 1 on 2, 3 accordingly.

It is necessary to mention that the algorithm (1.24)–(1.29) allows us to find the problem solution with boundary conditions of the 2nd or 3rd type on the right border of ranges of definition $R$. The representing equation (1.25) already comprises in advance initial (boundary) conditions on the left-side of the range of definition. For derivative findings on the right border we differentiate on $x_1$ expression (1.26) [it is supposed that there are limited private derivatives on $x_m, m = 1, 2, 3$ from $T(t, x)$]. Then we will receive, lowering an index 1/3 at $T(x)$:

$$\partial T(x)/\partial x_1 = \exp(-a_1 x_1) dg_1/dx_1 - A_1^{-1} \exp(-a_1 x_1) \times \int_0^{x_1} f(\xi, x_2, x_3) \exp(a_1 \xi)\, d\xi. \quad (1.30)$$

As a result at $x_1 = L_1$ we finally see

$$A_1 \partial T(L_1, x_2, x_3)/\partial x_1 = A_1 \frac{dg_1}{dx_1} \exp(-a_1 L_1) - \int_0^{L_1} f(\xi, x_2, x_3) \times \exp[-a_1(L_1 - \xi)]\, d\xi. \quad (1.31)$$

The condition (1.31) allows us again to find $dg_1/dx_1 (\partial T(L_1, x_2, x_3)/\partial x_1$ is considered to be known), and as a result expression of a type (1.28) of the first stage using now formulas (1.30) and (1.31). Similar conditions of types (1.30) and (1.31) can be received on other coordinates $x_2$ and $x_3$. To find the integrals entering into expressions (1.28) and (1.29), we use Simpson's formula [21].

Without loss of generality, we will check the solution (1.26) of Eq. (1.21) of the first stage for simplicity under zero boundary conditions: $g_1 = dg_1/dx_1 = q_1 = 0$. For this purpose, we differentiate the expression (1.27) on $x_1$. Then we find

$$\partial T^2/\partial x_1^2 = -A_1^{-1}[-a_1 \exp(-a_1 x_1) \int_0^{x_1} f(\xi, x_2, x_3) \\ \times \exp(a_1 \xi)\, d\xi + f(x)]. \quad (1.32)$$

As a result, substituting Eqs. (1.30) and (1.32) in the Eq. (1.21), we finally have

$$f + A_1 \partial T^2/\partial x_1^2 + B_1 \partial T/\partial x_1 = f + a_1 \exp(-a_1 x_1) \int_0^{x_1} f(\xi, x_2, x_3) \\ \times \exp(a_1 \xi)\, d\xi - f - a_1 \exp(-a_1 x_1) \int_0^{x_1} f(\xi, x_2, x_3) \exp(a_1 \xi)\, d\xi \equiv 0.$$

In the same way it is possible to check solutions (1.29) for the Eqs. (1.22) and (1.23) of the second and third stages, for example, at the zero boundary conditions.

## Examples of Test Calculation

Accuracy of the received analytical expressions (1.28) and (1.29) and correctness of calculation algorithm in a spatial case, we will establish the solution of three-dimensional differential problems (1.33) with [29] $c$ boundary conditions (1.34):

$$c_1 \sum_{m=1}^{3} \partial T/\partial x_m + c_2 \sum_{m=1}^{3} \partial^2 T/\partial x_m^2 + c_3 T + f(x) = 0; \quad (1.33)$$

$$f = -\left[ c_3 \left( c_4 + \sum_{m=1}^{3} x_m^z \right) + c_1 z \sum_{m=1}^{3} x_m^{z-1} + c_2 z(z-1) \sum_{m=1}^{3} x_m^{z-2} \right],$$

$$T|_{x_1=0} = (c_4 + x_2^z + x_3^z), \quad T|_{x_2=0} = (c_4 + x_1^z + x_3^z),$$
$$T|_{x_3=0} = (c_4 + x_1^z + x_2^z), \quad T|_{x_1=L_1} = (c_4 + L_1^z + x_2^z + x_3^z),$$
$$T|_{x_2=L_2} = (c_4 + L_2^z + x_1^z + x_3^z), \quad T|_{x_3=L_3} = (c_4 + L_3^z + x_1^z + x_2^z). \quad (1.34)$$

The exact solution of problems (1.33) and (1.34) in area: $0 \le x_m \le L_m$, $m = 1, 2, 3$ is known in advance: $T = c_4 + \sum_{m=1}^{3} x_m^z$. The following values of the entrance data for a basic variant were taken: $c_2 = c_4 = 1$, $z = 2, L_m = 1, N_m = 21, h_m = L_m/(N_m - 1), m = 1, 2, 3, N_m$ is the number of checkouts on space, $h_m$ are steps on the spatial variables at calculation of integrals in the Eqs. (1.28) and (1.29) with Simpson's formula [30]. The program of calculation of a boundary problem (1.33) and (1.34)

**Table 1.3** A dependence of the maximum relative error from various values of $c_1$ and $c_3$

| $N$ | 1 | 2 | 3 | 4 | 5 | 6 |
|---|---|---|---|---|---|---|
| $c_3$ | 1 | −1 | −1 | 1 | 0 | 0 |
| $c_1$ | 1 | 1 | −1 | −1 | 1 | −1 |
| $\varepsilon$, % | 0.38 | 0.23 | 0.05 | 0.17 | 0.15 | 0.092 |

with formulas (1.28) and (1.29) is made in Fortran-90 language, calculations are made on COMPUTER Pentium 2 (compiler PS 4, 130 MHz) with double accuracy.

Table 1.3 gives maximum relative error $\varepsilon = |T(x) - \tilde{T}(x)|/T(x)$ ($T(x)$ is accurate, $\tilde{T}(x)$ is the analytical solution) for different values $c_1$ and $c_3$ for the reference variants, where for $c_3 = 0$ is taken $\sigma = 1$ in Eq. (1.29). At the same time, calculation amounts ($t_0$): $t_0 = 2$ s.

It is of interest to obtain the solution of Eqs. (1.33) and (1.34) for real values of the input data, such as KKT in the problem of modeling the thermal protection of [31], where geometrical dimensions of the problem domain (the thickness of heat insulation coverage) is much less than 1 m; in this case $L_m = 10^{-2}$ m, $c_1 = 1$ m/s, $c_2 = 10^{-4}\,\text{m}^2/\text{s}$ (material such as copper [31]) and $c_4 = 10^{-2}$, $z = 2$, we have $\varepsilon = 2.4 \cdot 10^{-3}$.

Now compare the solution from Table 1.3 for the number $N = 5$ ($c_1 = 1$, $c_3 = 0$) of Eqs. (1.33) and (1.34) with numerical calculation by technology from [29] for the reference version. To achieve equal accuracy $\varepsilon = 0.15$ it is necessary to establish the method of [29] for $T_H = 1$ to 13 iterations ($t_0 = 5$ s), and at $T_H = 0$–64 iterations ($t_0 = 20$ s).

In order to verify the accuracy and effectiveness of the formulas (1.28) and (1.29) we have also considered the solution of the Dirichlet problem for the cube Poisson's equation (1.35) obtained from Eq. (1.33) with $c_1 \ll 1 (c_1 = 10^{-10}), c_2 = 1, c_3 = 0$.

$$\sum_{m=1}^{3} \partial^2 T/\partial x_m^2 + f(x) = 0,\ T\big|_{\Gamma} = 0, \quad x_m \in (0,1), \tag{1.35}$$

where $\Gamma$ is the boundary of the cube.

A source in the Eq. (1.35) as in Eq. (1.33), was selected in accordance with the condition that the test function $T = \prod_{m=1}^{3} x_m(1 - x_m)$ was the exact solution. We should note that in the case of two-dimensional formulation, the problem (1.35) coincides with the test task, which is given in the spheres of social [23] (with $c_2$ = const). The result is $\varepsilon = 0.015$ obtained from Eqs. (1.28) and (1.29) with $\sigma = 0.1$, and calculation time was $t_0 = 2$ s. According to the technology of [29] to obtain the same accuracy $\varepsilon = 0.015$ for other equal input data and $T_H = 0$ it is necessary to make 65 iterations and $T_0 = 10$ s (up to five times longer).

# CHAPTER 2

# Method of Solution of Nonlinear Transfer Equations

In mathematical modeling of problems of heat- and mass-transfer [1, 2] there is a problem with solution of mass conservation equations, momentum equations, and energy equations. A typical equation can be an equation in partial derivatives of the first order with variables in factors and nonlinear sources.

In this chapter on the basis of method quasi-linearization [7], operation calculation [6], and locally-one-dimensional splittings on spatial coordinates [12], analytical formulas while solving equations in partial derivatives of the first order have been found. For one-dimensional and three-dimensional nonlinear transfer equations the condition of unequivocal resolvability of boundary problems has been found and the estimation of speed of convergence of the iterative process has been given. There are test calculations of modeling boundary problems on the basis of trial functions and comparison results with a numerical method.

## 2.1 METHOD OF SOLUTION OF ONE-DIMENSIONAL NONLINEAR TRANSFER EQUATIONS

Let's give the algorithm method quasi-linearization [7] as an example of a solution of a boundary problem for a one-dimensional nonlinear equation in partial derivatives of the first order.

### Statement of a Problem and a Method Algorithm

Let us attempt to find a solution of transfer equation [14, 20]:

$$A_2 \frac{\partial u}{\partial t} + A_1 \frac{\partial u w(u)}{\partial x} = E(u, x, t), \quad u \geq 0 \tag{2.1}$$

in the area $\overline{Q_t} = \overline{Q} \times [0 < t \leq t_k]$, $\overline{Q} = Q + \Gamma$, $\Gamma$ is a border, $Q = (0 < x < b, 0 < b < \infty)$ with the initial condition

$$u\,|_{t=0} = \alpha(x). \tag{2.2}$$


http://dx.doi.org/10.1016/B978-0-12-804289-2.00002-8

Let's assume for definiteness that signs of sizes $w, A_p, p = 1, 2$ are known in advance, for example: $A_p > 0$, $w > 0$ at $0 \leq t \leq t_k$. Then a boundary condition in the region $\overline{Q_t}$ is performed in the form of [20]

$$u\,|_{x=0} = \gamma(t). \tag{2.3}$$

Let's assume in each case:

1. A problem (2.1)–(2.3) has the unique solution $u(x, t)$, which is continuous in the closed region $\overline{Q_t} = \overline{Q} \times [0 \leq t \leq t_k]$ and has continuous derivatives $\partial u/\partial t$, $\partial u/\partial x$.
2. The following conditions are provided: $\gamma \geq 0, \alpha \geq 0, E(x, t, u) \geq 0$ and only the positive solution of the boundary problem is found (2.1)–(2.3).
3. Functions $\alpha$, $\gamma$, $E(x, t, u)$ are continuous in the considered region.

The equations of a type (2.1) are applied in the mechanics of inert and reacting environments [1, 2].

Let's use Kirchhoff's transformation in [5, p. 112]:

$$v = \int_{u_H}^{u} w(u)\, du. \tag{2.4}$$

Then, taking into account the relations [5]

$$\frac{\partial w}{\partial x} = \frac{\partial w}{\partial u}\frac{\partial u}{\partial x}, \quad \frac{\partial v}{\partial t} = w\frac{\partial u}{\partial t}, \quad \frac{\partial v}{\partial x} = w\frac{\partial u}{\partial x}, \tag{2.5}$$

we have the differential equation from Eqs. (2.1) and (2.5)

$$\frac{A_2}{w}\frac{\partial v}{\partial t} + A_1\left(\frac{u}{w}\frac{\partial w}{\partial u} + 1\right)\frac{\partial v}{\partial x} = E(u, x, t). \tag{2.6}$$

Further, to use the inversion formula we take $w = u^m (m > 0)$ and we set $E(u, x, t) = A_3 u^k + A_4 \exp(A_5 u) + F(x, t), A_p = \text{const}, p = 3, 4, 5$ then from Eq. (2.4) we have [32]:

$$\begin{aligned} v &= [u^{m+1} - \alpha^{m+1}(x)]s, \quad s = (m+1)^{-1}, \\ u &= [v(m+1) + \alpha^{m+1}(x)]^s, \quad u_H = \alpha(x). \end{aligned} \tag{2.7}$$

Let's note that the variation ranges of independent variables and type of boundary conditions do not change in relation to Kirchhoff's transformation (2.4) and within the inversion formula (2.7) a boundary condition of the 1st type passes into Dirichlet's boundary condition. As a result from Eqs. (2.7), (2.2), and (2.3) initial and boundary conditions for the Eq. (2.6) will be

$$v\mid_{t=0} = v_H, \quad v_H = 0,$$
$$v\mid_{x=0} = s[\gamma^{m+1}(t) - \alpha^{m+1}(0)] = g(t). \tag{2.8}$$

Let $v_0$ = const be some initial approximation [as an initial approximation it is better to take the value close to $v_H$ from Eq. (2.8)]. Let's consider the sequence $\{v_n(t,x)\}$, defined by the recurrence relationship [7] (a point above means a partial derivative on time):

$$\frac{\partial v_{n+1}}{\partial x} = f + (v_{n+1} - v_n)\frac{\partial f}{\partial v_n} + (\dot{v}_{n+1} - \dot{v}_n)\frac{\partial f}{\partial \dot{v}_n}, \quad f = f(v_n, \dot{v}_n, x, t); \tag{2.9}$$

$$v_{n+1}(t,0) = g(t), \quad v_H = v_n(x), \quad n = 0, 1, 2, \ldots. \tag{2.10}$$

To reduce further records we will introduce the following notations:

$$a_2 = A_2/w, \quad a_1 = A_1(m+1), \quad f = (E - a_2\dot{v})/a_1,$$
$$c = v(m+1) + \alpha^{m+1}(x), \quad a_3 = c^s, \quad \partial u/\partial v = a_3/c,$$
$$q(v) = [A_3 k a_3^k + A_4 a_3 A_5 \exp(A_5 a_3)]/c, \quad q(v) = \frac{\partial f}{\partial v},$$
$$\bar{F} = F/a_1, \quad W(v) = \{A_3 a_3^k[1 - kv/c] + A_4 \exp(A_5 a_3)$$
$$\times [1 - A_5 a_3 v/c]\}/a_1, \frac{\partial f}{\partial \dot{v}} = -a_2/a_1, \quad a = q/a_1. \tag{2.11}$$

Each function $v_{n+1}(t,x)$ is the solution of linear equation, which is a rather important feature of this algorithm. The algorithm comes from a Newton-Kantorovich's approximation method [8] in functional space. We will take the solution of the linear differential problem (2.9) and (2.10). Let's substitute relationships (2.11) in the Eq. (2.9) and for simplicity of calculations we will introduce the right member of Eq. (2.9) through $h$, then it will be at $a$ from Eq. (2.11), obviously independent on $x$ [$a$ is always possible to set on the bottom iteration on $n$ knowing values $v_n$ in initial and subsequent moment of time from Eqs. (2.7) and (2.8)]:

$$\frac{\partial v_{n+1}}{\partial x} - a v_{n+1} = h(v_n, \dot{v}_{n+1}, x, t), \quad n = 0, 1, 2, \ldots. \tag{2.12}$$

Let's apply the Laplace integral transformation to the differential equation (2.12) excluding derivative on $x$ and replacing it with its linear expression concerning the image required function for which Laplace integral transformation converges absolutely. The valid part of complex number

$p = \xi + I\eta, I = \sqrt{-1}$ is considered positive, that is $\mathrm{Re}\, p > 0$. Then, using formulas from [6] and introducing letters $V, H$ for images, we will have:

$$pV_{n+1} - aV_{n+1} = v_{n+1}(t,0) + H(t,p)$$
$$V_{n+1} = v_{n+1}(t,0)/(p-a) + H(t,p)/(p-a). \tag{2.13}$$

Using the Laplace integral transformation from [6]: $L^{-1}[H(p)/p] = \int_0^x h(y)\, dy$, $L^{-1}[(p-a)^{-1}] = \exp(ax)$, we restore the original for $v_{n+1}(x,t)$ from Eq. (2.13):

$$v_{n+1}(x,t) = v_{n+1}(t,0)\exp(ax) + \int_0^x \exp[a(x-y)] \times h(v_n, \dot{v}_{n+1}, y, t)\, dy, \quad n = 0,1,2,\ldots. \tag{2.14}$$

Using relationships (2.11) again, the Eq. (2.14) can be rewritten in a detailed form:

$$\dot{v}_{n+1} + Uv_{n+1} = Z^{-1}\left\{v_{n+1}(t,0)\exp(ax) + \int_0^x \exp[a(x-y)] \times (\bar{F} + W)\, dy\right\} = Y(v_n, x, t), \quad v_n(0,x) = v_H,$$
$$n = 0,1,2,\ldots,$$
$$Z = \frac{A_2}{A_1(m+1)}\int_0^x \frac{\exp[a(x-y)]}{a_3^m}\, dy, \quad U = U(v_n) = Z^{-1}. \tag{2.15}$$

As a result the problem final solution (2.15) will be at $U$, obviously independent on $t$ according to [26]:

$$v_{n+1}(t,x) = \left[v_H + \int_0^t Y(v_n, x, \tau)\exp(U\tau)\, d\tau\right]\exp(-Ut), n = 0,1,2,\ldots. \tag{2.16}$$

As a result, the solution $u(t,x)$ of a nonlinear boundary problem (2.1)–(2.3) is found from the inversion formula (2.7) at $v(t,x)$, defined from Eq. (2.16).

## Existence, Uniqueness, and Convergence

We will consider a nonlinear case [32]: $m \geq 1, k > 1, w = u^m$ and for simplicity of the analysis at $A_p = 1, p = 1,2,3, A_4 = F = 0$.

Then the solution of a boundary problem

$$\frac{\partial u^{m+1}}{\partial x} = u^k - \dot{u}, \quad u(0,x) = \alpha(x), \quad u(0,t) = \gamma(t) \tag{2.17}$$

using the algorithm (2.7)–(2.16) can be written as at $g(t) = 0$;

$$u = [v(m+1) + \alpha^{m+1}(x)]^s, \quad v_n(x,0) = v_H(x),$$

$$v_{n+1} = \left\{ v_H + \int_0^t Z^{-1}(v_n) \left[ \int_0^x \exp[a(x-y)] W(y)\, dy \right] \times \exp(U\tau)\, d\tau \right\} \exp(-Ut), \quad n = 0, 1, 2, \ldots,$$

$$Z(v_n) = \frac{1}{(m+1)} \int_0^x \frac{\exp[a(x-y)]}{a_3^m} dy, \quad U = U(v_n) = Z^{-1}(v_n). \quad (2.18)$$

For simplicity of further calculations we will introduce the following notation:

$$f(v_n) = [Z(v_n)]^{-1} \int_0^x \exp[a(x-y)] W\, dy \cdot \exp[U(v_n)\tau]$$

then $v_{n+1}$ from Eq. (2.18) will be rewritten as:

$$v_{n+1} = \left[ v_H + \int_0^t f(v_n)\, d\tau \right] \exp[-tU(v_n)], \quad n = 0, 1, 2, \ldots \quad (2.19)$$

In order to use a Lipschitz condition [30] we will assume that $f(v)$ from Eq. (2.19) has in the considered region $\overline{Q_t} : [0 \le x \le b, 0 \le t \le t_k]$ the limited partial derivative $\partial f(v)/\partial v$.

**Theorem.** *Let* $v$ *be continuously differentiated in* $\overline{Q_t}$, *then in the region* $\overline{Q_t}$ *there is a unique solution of a boundary problem* (2.17).

## Existence

We will prove that the sequence of approximation set by equalities (2.18) converges in regular intervals to some function $v \in R$, for example in the closed region $\overline{Q_t}$. Then in equality (2.18) it will be possible to pass under the badge of the integral to a limit $n \to \infty$, and we have:

$$v = \left\{ v_H + \int_0^t Z^{-1}(v) \left[ \int_0^x \exp[a(x-y)] W(y)\, dy \right] \times \exp(U(v)\tau) d\tau \right\} \exp(-U(v)t). \quad (2.20)$$

Differentiation on $t$ (2.20) gives the equation

$$\dot{v} + Uv = [Z(v)]^{-1} \int_0^x W \exp[a(x-y)]\, dy, \quad v(0,x) = v_H. \quad (2.21)$$

Value $v$ at $t = 0$ from Eq. (2.20) also gives the condition for $v(0, x)$.

To prove convergence of sequence $v_n$, we will consider a row

$$\sum_{n=0}^{n=\infty} (v_{n+1} - v_n). \tag{2.22}$$

The sequence $v_n$ will converge in regular intervals if a row (2.22) converges in regular intervals. But a row (2.22) converges in regular intervals, if a majorized row $\sum_{n=0}^{n=\infty} |v_{n+1} - v_n|$ converges in regular intervals. From recurrence relationship (2.19) we have:

$$v_n = \left[ v_H + \int_0^t f(v_{n-1})\, d\tau \right] \exp[-tU(v_{n-1})], \quad n \geq 1. \tag{2.23}$$

Let $U(v_n) \geq 0$, therefore $\max\limits_{v_n \in R}\{\exp[-tU(v_n)]\} \leq 1, n = 0, 1, 2, \ldots$. As a result, subtracting from the Eq. (2.19) expression (2.23), we find

$$|v_{n+1} - v_n| \leq \max_{v_n \in R} \int_0^t |f(v_n) - f(v_{n-1})|\, d\tau, \quad n \geq 1. \tag{2.24}$$

Let's assume further that for any $v, u \in R$ the Lipschitz condition is provided

$$|f(u) - f(v)| \leq c_1 |u - v|, \tag{2.25}$$

where $c_1$ is the constant, which is independent on functions $u, v$. Coming back to an inequality (2.24) and using Eq. (2.25), we have:

$$\| v_{n+1} - v_n| \leq c_1 \max_{v_n \in R} \int_0^t |v_n - v_{n-1}|\, d\tau, \quad n \geq 1. \tag{2.26}$$

As from a relationship (2.24): $|v_1 - v_0| \leq \max\limits_R \int_0^t |f(v_0)|\, d\tau = \max\limits_R |f(v_0)| t = c_2 t$ $(c_2 = \max\limits_R |f(v_0)|)$, with the iteration procedure in the formula (2.26) $|v_2 - v_1| = c_1 \int_0^t |v_1 - v_0|\, d\tau \leq c_2 c_1 t^2/2, \ldots$, we come to an inequality

$$\| v_{n+1} - v_n| \leq c_2 c_1^n t^{n+1}/(n+1)!. \tag{2.27}$$

Uniform convergence of a row for exponential function from Eq. (2.27) on each final interval provides uniform convergence of a row $\sum_{n=0}^{n=\infty} |v_{n+1} - v_n|$, and therefore uniform convergence of sequence $|v_n|$ to function $v$, satisfying the Eq. (2.20), and to an initial problem (2.21), and as a result of the regional (initial-boundary) differential problem (2.17) according to the inversion formula (2.7).

## Uniqueness

We will show that the problem solution (2.17) is the only thing in the considered region $\overline{Q_t}$. We will admit that there is another solution $u$ of this problem. As $u$ is continuous and can be performed as

$$u = \left( v_H + \int_0^t f(u)\, d\tau \right) \exp[-tU(u)] \tag{2.28}$$

and its value at $t = 0$ contains in $R$, $u \in R$ at $[0 \le x \le b, 0 < t \le t_k]$. Then, combining equality (2.28) (it is supposed, as well as above that $U(u) \ge 0$) with the formula (2.19), similarly Eqs. (2.24)–(2.26) we have:

$$|v_{n+1} - u| \le \max_R \int_0^t |f(v_n) - f(u)|\, d\tau,$$

$$|v_{n+1} - u| \le c_3 \max_R \int_0^t |v_n - u|\, d\tau. \tag{2.29}$$

As $|v_0 - u| \le |v_0| + |u| \le c_4, c_4 = |v_H| + c_5, c_5 = \max_R |u|$. Taking into account that from an inequality (2.29) it follows that $|v_1 - u| \le c_3 \max_R \int_0^t |v_0 - u|\, d\tau \le c_3 c_4 t$, we find with the help of iterations $|v_2 - u| \le c_3 \int_0^t |v_1 - u|\, d\tau \le c_4 (c_3 t)^2/2, \ldots$ an inequality

$$\| v_{n+1} - u| \le c_4 (c_3 t)^{n+1}/(n+1)!. \tag{2.30}$$

Going to Eq. (2.30) $n \to \infty$, we have $|v - u| \le 0$, where $v \equiv u$.

## Estimation of Speed of Convergence [32]

It is supposed that in some neighborhood of a root function $f = f(v, \dot{v})$ from Eq. (2.9) together with the partial derivatives $\partial f/\partial v, \partial^2 f/\partial v^2, \partial f/\partial \dot{v}, \partial^2 f/\partial \dot{v}^2$ are continuous, and $\partial f/\partial v, \partial^2 f/\partial v^2, \partial f/\partial \dot{v}, \partial^2 f/\partial \dot{v}^2$ in this neighborhood do not go to zero.

Let's address the recurrence relationship (2.9) and, noticing that $f(v, \dot{v}) = s(v) - r(\dot{v})$ from Eq. (2.17), we will subtract $n$-e equation from $(n+1)$th then we will find:

$$\begin{aligned} \partial(v_{n+1} - v_n)/\partial x = {} & s(v_n) - s(v_{n-1}) - (v_n - v_{n-1}) \frac{\partial s(v_{n-1})}{\partial v} \\ & + (v_{n+1} - v_n) \frac{\partial s(v_n)}{\partial v} - [r(\dot{v}_n) - r(\dot{v}_{n-1}) \\ & - (\dot{v}_n - \dot{v}_{n-1}) \frac{\partial r(\dot{v}_{n-1})}{\partial \dot{v}} + (\dot{v}_{n+1} - \dot{v}_n) \frac{\partial r(\dot{v}_n)}{\partial \dot{v}} \Bigg]. \end{aligned} \tag{2.31}$$

From the average theorem [33] it follows:

$$s(v_n) - s(v_{n-1}) - (v_n - v_{n-1})\frac{\partial s(v_{n-1})}{\partial v} = 0.5(v_n - v_{n-1})^2 \times \partial^2 s(\xi)/\partial v^2, v_{n-1} \leq \xi \leq v_n.$$

Let's see Eq. (2.31) how the equation is in relation to $u_{n+1} = v_{n+1} - v_n$. We will transform as above Eqs. (2.7)–(2.15). Then we will have at $a = \frac{\partial s(v_n)}{\partial v}$:

$$u_{n+1} = \int_0^x \exp[a(x-y)] \left\{ 0.5 \left[ -\dot{u}_n^2 \frac{\partial^2 r(\dot{v}_n)}{\partial \dot{v}^2} + u_n^2 \frac{\partial^2 s(v_n)}{\partial v^2} \right] - \dot{u}_{n+1} \frac{\partial r(\dot{v}_n)}{\partial \dot{v}} \right\} dy$$

or

$$\dot{u}_{n+1} + U u_{n+1} = \frac{Z^{-1}}{2} \int_0^x \exp[a(x-y)] \left[ -\dot{u}_n^2 \frac{\partial^2 r(\dot{v}_n)}{\partial \dot{v}^2} + u_n^2 \frac{\partial^2 s(v_n)}{\partial v^2} \right] dy, \quad U = Z^{-1}, \tag{2.32}$$

$$Z = \int_0^x \exp[a(x-y)] \frac{\partial r(\dot{v}_n)}{\partial \dot{v}} dy.$$

As a result the problem solution (2.32) will look like Eq. (2.16), where $v_H = 0$. Let's put $\max\limits_{v,\dot{v}\in R} \left( \left| \frac{\partial s(v)}{\partial v} \right|, \left| \frac{\partial r(\dot{v})}{\partial \dot{v}} \right| \right) = c_1$, $\max\limits_{v,\dot{v}\in R} (|\partial^2 s(v)/\partial v^2|, |\partial^2 r(\dot{v})/\partial \dot{v}^2|) = c_2$, assuming that $c_m < \infty$, $m = 1, 2$.

Then from the Eqs. (2.16) and (2.32) it follows that $\partial^2 r(\dot{v})/\partial \dot{v}^2 = 0$:

$$\|u_{n+1}| \leq B \int_0^t u_n^2 \exp[-U(t-\tau)]\, d\tau. \tag{2.33}$$

Let's choose $u_0(t,x)$ so that $|u_0(x,t)| \leq 1$ in area Q. Hence, providing that $U > 0$ : $\max\limits_{0\leq\tau\leq t} \exp[-U(t-\tau)] \leq 1$ from the Eq. (2.33) we receive with $n = 0$, introducing $M_1 = \max\limits_{Q} |u_1|$, $B = c_2/2c_1$, $Z = \exp(c_1 b) - 1$, $U = Z^{-1}$:

$$M_1 \leq Bt = Y_1. \tag{2.34}$$

Then the top border $M_1$ will not surpass 1, if the inequality $Y_1 \leq 1$ in Eq. (2.34) is:

$$t \leq 1/B. \tag{2.35}$$

Therefore, if choosing an interval $[0, t]$ small enough, so that the condition was provided from Eq. (2.35), we will have $M_1 \leq 1$. Going on we finally receive $M_{n+1} \leq Y_1 u_n^2$ or

$$\max_{x,t \in Q} |v_{n+1} - v_n| \leq Y_1 \left[ \max_{x,t \in Q} |v_n - v_{n-1}|^2 \right]. \tag{2.36}$$

The relationship (2.36) shows that if convergence in general takes place, it is quadratic. The result is that each following step doubles the number of correct signs in the given approximation.

## Example of Test Calculation

We check the accuracy of the received approximate analytical formula (2.16) practically while solving the nonlinear boundary problem for the equation in partial derivatives:

$$A_2 \frac{\partial u}{\partial t} + A_1 \frac{\partial u^{m+1}}{\partial x} = A_3 u^k + A_4 \exp(A_5 u) + E(x, t); \tag{2.37}$$

$$u \mid_{x=0} = \exp(t); \tag{2.38}$$

$$u \mid_{t=0} = \exp(x/b). \tag{2.39}$$

If the exact solution (2.37)–(2.39) is taken as: $u = \exp(t + y), y = x/b$, a source $E$ in the Eq. (2.37) will be

$$\begin{aligned} E = \exp(t + y)\{A_2 + A_1 b^{-1}(m + 1)\exp[m(t + y)]\} \\ - A_3 \exp[k(t + y)] - A_4 \exp[A_5 \exp(y + t)]. \end{aligned}$$

We have taken the following basic values of the initial data: $m = 1, A_1 = A_2 = 1, b = 1, t = 1, N = 11, M = 51, h = b/(N - 1), \tau = t/(M - 1)$; $N, M, h, \tau$ is a number of checkouts and steps on space and on time while finding integrals in the Eqs. (2.15) and (2.16) by trapezoid formula and Simpson's formula [21]. The program is made on Fortran-90, calculation was made on Pentium 2 (the compiler PS 4, 130 MHz) with double accuracy.

In Table 2.1 there are the results of calculations of the maximum relative errors $\varepsilon = |u - \tilde{u}| \cdot 100\%/u$ ($u$ is exact, $\tilde{u}$ is approximate analytical solution) at various values $A_p, p = 3, 4, 5$ and $k$ for a basic variant.

The boundary problem (2.37)–(2.39) was solved by method quasi-linearization by means of substitution (2.7) and formulas (2.16). Number of iterations ($J$) was traced on relative change of vector errors (in percentage):

**Table 2.1** A dependence of the maximum relative error at the various values $A_p, p = 3, 4, 5$, and $\kappa$

| N | 1 | 2 | 3 | 4 | 5 | 6 | 7 |
|---|---|---|---|---|---|---|---|
| $\kappa$ | 1 | 1 | 2 | 2 | 1 | 1 | 1 |
| $A_3$ | 0 | −1 | −1 | 1 | 1 | 0 | 1 |
| $A_4$ | 0 | 0 | 0 | 0 | 0 | 1 | 1 |
| $A_5$ | 0 | 0 | 0 | 0 | 0 | −1 | −1 |
| $\varepsilon$, % | 2.21 | 2.27 | 2.14 | 2.55 | 2.13 | 2.22 | 2.15 |
| $J$ | 3 | 3 | 3 | 3 | 3 | 3 | 3 |

$$||u_n|| = \max_{x,t \in Q} |(v_{n+1} - v_n)/v_{n+1}|.$$

Table 2.1 shows the results of calculations by $J$ with $||u_n|| \le \delta$, $\delta = 1\%$. Thus time of calculation of any variant = 1 s.

Let's compare the problem solution (2.37)–(2.39) on relationships (2.7) and (2.16) with numerical calculation [20]. As the exact solution of a problem (2.37)–(2.39) was given, let's use the implicit absolutely steady formula of the central differences on space [20] for internal nodes at $A_4 = A_5 = 0$:

$$\begin{aligned} u_0^{(j+1)} &= \exp(t^{(j+1)}), -u_{n+1,i-1}^{(j+1)} q_{i-1}/2h + u_{n+1,i}^{(j+1)}(\tau^{-1} - A_3 g_i) \\ &\quad + u_{n+1,i+1}^{(j+1)} q_{i+1}/2h \\ &= (u\tau^{-1} + E)_i^{(j)}, \quad i = \overline{1, N-1}, \quad n = 0, 1, 2, \ldots, \end{aligned} \tag{2.40}$$

$$q = u_n^j, \quad g = (u_n^{k-1})^{(j)}, \quad n = 0, 1, 2, \ldots,$$

and the last equation in the node $x = x_N$ for resolution of three-pointed run taken from [34], represents approximation of the initial Eq. (2.37) [35] and has the second order of accuracy on space on the problem solution:

$$\begin{aligned} &u_{n+1,N}^{(j+1)}(\tau^{-1} + 2q_N/h - A_3 g_N) + u_{n+1,N-1}^{j+1}(\tau^{-1} - 2q_{N-1}/H \\ &\quad - A_3 g_{N-1}) = (u\tau^{-1} + E)_N^{(j)} + (u\tau^{-1} + E)_{N-1}^{(j)}, \quad n = 0, 1, 2, \ldots . \end{aligned} \tag{2.41}$$

Approximation error of difference schemes (2.40) and (2.41) on time—$O(\tau)$, and difference equations (2.40) on space—$O(h^2)$. Last estimation coincides with accuracy of calculation of integrals in the ratio (2.15) by the trapezoid formula of even number of nodes on space. At numerical realization of calculation $u_{n+1,i}^{(j+1)}$, $n = 0, 1, 2, \ldots$, $i = \overline{1, N}$ on each step of time, the Pikar's method of consecutive approximations [30] which converged for two iterations was used.

For $\tau = 0.02, k = 1, A_3 = -1$ and the basic initial data calculation on relationships (2.40) and (2.41) gives $\varepsilon = 9.8\%$ by the time of time $t = 1$, and in Table 2.1 at number 2 we have: $\varepsilon = 2.27\%$. However at $A_3 = -1$ and $k = 2$ by formulas (2.40) and (2.41) it turns out to be $\varepsilon =$ 39%, and in Table 2.1 at number 3 is $\varepsilon = 2.14\%$, which is more exact by an order.

## 2.2 ALGORITHM OF SOLUTION OF THREE-DIMENSIONAL NONLINEAR TRANSFER EQUATIONS

### Statement of a Problem and a Method Algorithm

Let us attempt to find the solution of a nonlinear transfer equation [1, 2, 14]

$$A_4\frac{\partial u}{\partial t} + \sum_{j=1}^{3} A_j\frac{\partial uw(u)}{\partial x_j} = E(u, x, t), \quad u \geq 0 \tag{2.42}$$

in the region $\overline{Q_t} = \overline{Q} \times (0 < t \leq t_k, 0 < t_k < \infty)$, $\overline{Q} = Q + \Gamma$, $Q = (0 < x_j < S_j, 0 < S_j < \infty, j = 1, 2, 3)$; $\Gamma$ is a boundary surface with initial condition

$$u|_{t=0} = \alpha(x). \tag{2.43}$$

Let's assume for definiteness that signs on sizes $w, A_j, j = \overline{1, 4}$ are known in advance, for example, $A_j > 0, w > 0$ at $0 \leq t < \infty$. Then boundary conditions in the region $\overline{Q_t}$ are established in the form of [1, 16]

$$u|_{x_1=0} = \beta_1(t, x_2, x_3), \quad u|_{x_2=0} = \beta_2(t, x_1, x_3), \quad u|_{x_3=0} = \beta_3(t, x_1, x_2). \tag{2.44}$$

Let's assume in each case:

1. A problem (2.42)–(2.44) has the unique solution $u(x, t)$, which is continuous in the closed region $\overline{Q_t} = \overline{Q} \times [0 \leq t \leq t_k]$ and has continuous derivatives $\partial u/\partial t$, $\partial u/\partial x_j$, $j = 1, 2, 3$.
2. The following conditions are provided: $\beta_m \geq 0, m = 1, 2, 3, \alpha \geq 0$, $E(x, t, u) \geq 0$ and the positive solution of a boundary problem is the only possible one (2.42)–(2.44).
3. Functions $\alpha$, $\beta_m$, $m = 1, 2, 3$, $E(x, t, u)$ are continuous in the considered region.

The equations of a type (2.42) is used in the mechanics of continuous environments [1, 2]. We will use Kirchhoff's transformation [5]

$$v = \int_{u_H}^{u} w(u)\, du. \tag{2.45}$$

We use relationships [5]

$$\nabla w = \frac{\partial w}{\partial u}\nabla u, \quad \frac{\partial v}{\partial t} = w\frac{\partial u}{\partial t}, \quad \nabla v = w\nabla u. \tag{2.46}$$

We receive the differential equation from Eqs. (2.42) and (2.46)

$$\frac{A_4}{w}\frac{\partial v}{\partial t} + \sum_{j=1}^{3} A_j\left(\frac{u}{w}\frac{\partial w}{\partial u} + 1\right)\frac{\partial v}{\partial x_j} = E(u, x, t). \tag{2.47}$$

Using the inversion formula, we will assume in each case $w = u^m (m > 0)$ and we will establish $E(u, x, t) = A_5 u^k + A_6 \exp(A_7 u) + F(x, t), A_j =$ const, $j = \overline{1, 7}, j \neq 4$. Then from Eq. (2.45) we have

$$v = [u^{m+1} - \alpha^{m+1}(x)] \cdot \gamma, \quad \gamma = (m+1)^{-1},$$
$$u = [v(m+1) + \alpha^{m+1}(x)]^{\gamma}, \quad u_H = \alpha(x). \tag{2.48}$$

As a result from Eqs. (2.43), (2.44), and (2.48) initial and boundary conditions for the Eq. (2.47) will be

$$v|_{t=0} = v_H, \quad v_H = 0, \quad v|_{x_1=0} = \gamma[\beta_1^{m+1} - \alpha^{m+1}(x_2, x_3)] = g_1(t, x_2, x_3),$$
$$v|_{x_2=0} = \gamma[\beta_2^{m+1} - \alpha^{m+1}(x_1, x_3)] = g_2(t, x_1, x_3),$$
$$v|_{x_3=0} = \gamma[\beta_3^{m+1} - \alpha^{m+1}(x_1, x_2)] = g_3(t, x_1, x_2). \tag{2.49}$$

The idea of a method of multidimensional problem reduction to sequence one-dimensional problems [12] we will use as for three-dimensional linear transfer equations in the absence of a source:

$$\frac{\partial v}{\partial t} + \sum_{j=1}^{3} c_j \frac{\partial v}{\partial x_j} = 0, \quad 0 < x_j \leq S_j,$$
$$c_j > 0, \quad c_j = \text{const}, \quad j = 1, 2, 3; \tag{2.50}$$

$$v|_{t=0} = \psi(x), \quad v|_{x_1=0} = g_1, \quad v|_{x_2=0} = g_2, \quad v|_{x_3=0} = g_3. \tag{2.51}$$

Let's consider locally-one-dimensional scheme splittings and introduce the following notations $R_j v_{(j)} = c_j \frac{\partial v}{\partial x_j}, j = 1, 2, 3$, then we have [12]:

$$\frac{\partial v_{(1)}}{\partial t} + R_1 v_{(1)} = 0, \quad v_{(1)}(0, x) = \psi(x),$$
$$v_{(1)}|_{x_1=0} = g_1, \quad 0 < t < t_*; \tag{2.52}$$

$$\frac{\partial v_{(2)}}{\partial t} + R_2 v_{(2)} = 0, \quad v_{(2)}(0,x) = v_{(1)}(t_*, x),$$
$$v_{(2)}|_{x_2=0} = g_2, \quad 0 < t < t_*; \tag{2.53}$$
$$\frac{\partial v_{(3)}}{\partial t} + R_3 v_{(3)} = 0, \quad v_{(3)}(0,x) = v_{(2)}(t_*, x),$$
$$v_{(3)}|_{x_3=0} = g_3, \quad 0 < t < t_*. \tag{2.54}$$

According to this model the process of transfer "is stretched out" in time and occurs at a time interval $3t_*$, instead of $t_*$ [13]. Let's consider the two-dimensional Laplace integral transformation [6, p. 314]:

$$V(p, x_1) = \int_0^\infty \exp(-pt) v(t, x_1)\, dt,$$
$$V_1(p, q) = \int_0^\infty \exp(-qx_1) V(p, x_1)\, dx_1. \tag{2.55}$$

Here $p, q$ is complex parameters. Further we consider the functions for which Laplace integral transformation converges absolutely [6]. The valid parts of numbers $p$ and $q$ are considered as positive, that is Re $p > 0$, Re $q > 0$. Then the Laplace integral transformation to Eq. (2.55) gives [6]:

$$L_q^{-1}[V_1] = V(p, x_1), \quad L_p^{-1}[V] = v(t, x_1). \tag{2.56}$$

Let's use Laplace integral transformation (2.55) at first to the Eq. (2.52). We assume that desired solution $v_1(t,x)$, and also its derivatives in the Eq. (2.52), satisfy the conditions of Laplace integral transformation on $t$ and on $x_1$, and growth degree on $t$ functions $v_1(t,x)$ and its derivatives does not depend on $x_1$. Similarly growth degree on $x_1$ functions $v_1(t,x)$ and its derivatives does not depend on $t$. Multiplying both parts of the first equation (2.52) on $\exp(-pt)$ and integrating on $t$ from 0 to $\infty$, and then multiplying on $\exp(-qx_1)$ and integrating on $x_1$ from 0 to $\infty$, we will have (applying integration rule) [36]:

$$c_1 \frac{\partial V(p,x)}{\partial x_1} + pV(p,x) = v(0,x), \quad c_1 q V_1(p,q,x_2,x_3) - c_1 V(p,0,x_2,x_3)$$
$$+ pV_1(p,q,x_2,x_3) = V(0,q,x_2,x_3). \tag{2.57}$$

Let's transform the Eq. (2.57), collecting similar members at $V_1(p,q,x_2,x_3)$:

$$V_1(p,q,x_2,x_3) = [V(0,q,x_2,x_3) + c_1 V(p,0,x_2,x_3)]/(p+a), \quad a = c_1 q. \tag{2.58}$$

For finding originals in Eq. (2.58) we will use formulas from [6, p. 151]:

$$\eta(t)\exp(-p\tau) = \begin{cases} 0, & t < \tau,\ \tau \geq 0, \\ \eta(t-\tau), & t \geq \tau; \end{cases} \tag{2.59}$$

$$L_p^{-1}\left[\frac{V(p)}{p+a}\right] = \int_0^t \exp(-a\tau)v(t-\tau)\,d\tau,$$
$$L_p^{-1}[\exp(-p\tau)V(p)] = v(t-\tau), \quad \tau > 0. \tag{2.60}$$

Using the Laplace integral transformation (2.56) consecutively, for example, to the second summand of the right-hand side of Eq. (2.58), and using Eq. (2.60), we find:

$$L_p^{-1}\left[\frac{V(p,0,x_2,x_3)}{p+a}\right] = \int_0^t \exp(-\tau c_1 q)g_1(t-\tau,0,x_2,x_3)\,d\tau = S,$$
$$L_q^{-1}[S] = J_1, \tag{2.61}$$

where

$$J_1 = \int_0^t g_1(t-\tau, 0-\tau c_1, x_2, x_3)\,d\tau, \quad \tau c_1 \leq 0,$$
$$J_1 = 0, \quad \tau c_1 > 0.$$

Let's note that to receive the last expression $J_1$ in a chain (2.61) functional dependence (2.59) is used. It is similarly restored through Eqs. (2.56), (2.59), and (2.60) the original for another summand in Eq. (2.58). As a result the solution of the Eq. (2.52) is finally written as:

$$v_{(1)}(t_*, x) = \psi(0, x_1 - t_* c_1, x_2, x_3)$$
$$+ c_1 \int_0^{t_*} g_1(t_* - \tau, 0 - \tau c_1, x_2, x_3)\,d\tau. \tag{2.62}$$

Using algorithm (2.55)–(2.61), we will write the solution of other Eqs. (2.53) and (2.54):

$$v_{(2)}(t_*, x) = v_{(1)}(t_*, x_1, x_2 - t_* c_2, x_3) + c_2 \int_0^{t_*} g_2(t_* - \tau, x_1, 0\tau c_2, x_3)\,d\tau; \tag{2.63}$$

$$v_{(3)}(t_*, x) = v_{(2)}(t_*, x_1, x_2, x_3 - t_* c_3) + c_3 \int_0^{t_*} g_3(t_* - \tau, x_1, x_2, 0 - \tau c_3)\,d\tau. \tag{2.64}$$

Let's substitute $v_{(1)}$ from Eq. (2.62) in the Eq. (2.63) for $v_{(2)}$, and then $v_{(2)}$ from Eq. (2.65) in the Eq. (2.64) for $v_{(3)}$ then we finally receive the desired solution of a boundary problem (2.50) and (2.51):

$$v_{(2)}(t_*, x) = \psi(x_1 - t_*c_1, x_2 - t_*c_2, x_3) + c_1 \int_0^{t_*} g_1(t_* - \tau, 0 - \tau c_1, x_2 - \tau c_2, x_3)\, d\tau + c_2 \int_0^{t_*} g_2(t_* - \tau, x_1, 0 - \tau c_2, x_3)\, d\tau; \tag{2.65}$$

$$v_{(3)}(t_*, x) = \psi(x_1 - t_*c_1, x_2 - t_*c_2, x_3 - t_*c_3) + \sum_{j=1}^{3} c_j I_j = v(t_*, x), \quad \forall t_* > 0; \tag{2.66}$$

$$\begin{aligned}
I_1 &= \int_0^{t_*} g_1(t_* - \tau, 0 - \tau c_1, x_2 - \tau c_2, x_3 - \tau c_3)\, d\tau, \quad \tau c_1 \le 0,\\
I_1 &= 0, \quad \tau c_1 > 0;\\
I_2 &= \int_0^{t_*} g_2(t_* - \tau, x_1, 0 - \tau c_2, x_3 - \tau c_3)\, d\tau, \quad \tau c_2 \le 0,\\
I_2 &= 0, \quad \tau c_2 > 0;\\
I_3 &= \int_0^{t_*} g_3(t_* - \tau, x_1, x_2, 0 - \tau c_3)\, d\tau, \quad \tau c_3 \le 0,\\
I_3 &= 0, \quad \tau c_3 > 0.
\end{aligned} \tag{2.67}$$

Let's apply locally-one-dimensional scheme splitting to Eq. (2.47) at differential level [12]:

$$\frac{\partial v^{(1)}}{\partial x_1} = -a_1 \frac{\partial v^{(1)}}{\partial t} + \sigma_1 z_1, \quad 0 < t < t_*, \tag{2.68}$$

$$v^{(1)}(0, x) = v_H(x), \quad v^{(1)}|_{x_1=0} = g_1(t, x_2, x_3); \tag{2.69}$$

$$\frac{\partial v^{(2)}}{\partial x_2} = -a_2 \frac{\partial v^{(2)}}{\partial t} + \sigma_2 z_2 + a_5 (u^{(1)})^k, \quad 0 < t < t_*, \tag{2.70}$$

$$v^{(2)}(0, x) = v^{(1)}(t_*, x), \quad v^{(2)}|_{x_2=0} = g_2(t, x_1, x_3); \tag{2.71}$$

$$\frac{\partial v^{(3)}}{\partial x_3} = -a_3 \frac{\partial v^{(3)}}{\partial t} + \sigma_3 z_3 + a_6 \exp(A_7 u^{(2)}), \quad 0 < t < t_*, \tag{2.72}$$

$$v^{(3)}(0, x) = v^{(2)}(t_*, x), \quad v^{(3)}|_{x_3=0} = g_3(t, x_1, x_2), \tag{2.73}$$

where $a_j = \gamma A_4/(wA_j)$, $z_j = \gamma F(x,t)/A_j, j = 1,2,3, \sigma_1 + \sigma_2 + \sigma_3 = 1, a_5 = \gamma A_5/A_2, a_6 = \gamma A_6/A_3$.

Our purpose is to find the solution of a nonlinear boundary problem, if it exists, as a limit of sequence of solutions of linear boundary problems. For this purpose we will use results of work [7] and will further assume that all coordinate directions in spacing are equivalent.

Let $v_0 = \text{const}$ be an initial approximation [as an initial approximation it is better to take the value close to $v_H$ from Eq. (2.49)]. Let's consider for simplicity of the analysis the quasi-one-dimensional case and sequence $v_n(t,x)$, defining recurrence relationship [7] (the point corresponds to the partial derivative on time):

$$\frac{\partial v_{n+1}}{\partial y} = f + (v_{n+1} - v_n)\frac{\partial f}{\partial v_n} + (\dot{v}_{n+1} - \dot{v}_n)\frac{\partial f}{\partial \dot{v}_n},$$
$$f = f(v_n, \dot{v}_n, x, t); \tag{2.74}$$

$$v_H = v_n(0,x), \quad v_{n+1}|_{x_1=0} = g_1, \quad v_{n+1}|_{x_2=0} = g_2,$$
$$v_{n+1}|_{x_3=0} = g_3, \quad n = 0,1,2,\ldots, \tag{2.75}$$

where $y$ is any of coordinates $x_j, j = 1,2,3$ in Eq. (2.74). Then at $y = x_1$ other coordinates in Eq. (2.74), $0 < x_j \leq S_j, j = 2,3$ change parametrically. With remained coordinates while receiving expression (2.74) there is a circular replacement of indexes when $y$ can be substituted by $x_2, x_3$. We will see that in the solution of the three-dimensional boundary problem (2.47) and (2.49) when in the first coordinate direction $x_1$ the initial iteration acts as $v_n$, according to the algorithm (2.52)–(2.67) the subsequent iteration $v_{n+1}$ can be found out of a definite expression (2.66), in which it is necessary to put $v_{n+1}(t_*, x) = v^{(3)}(t_*, x)$. Then in the quasi-one-dimensional variant of the equation in Eqs. (2.74) and (2.75) the coordinate $x_1$ will be rewritten:

$$\frac{\partial v^{(1)}}{\partial x_1} = f_1 + (v^{(1)} - v^{(0)})\frac{\partial f_1}{\partial v^{(0)}} + (\dot{v}^{(1)} - \dot{v}^{(0)})\frac{\partial f_1}{\partial \dot{v}^{(0)}}, \tag{2.76}$$
$$v^{(0)} = v_n, \quad f_1 = f_1(v^{(0)}, \dot{v}^{(0)});$$
$$v_H^{(1)} = v_H(x), \quad v^{(1)}|_{x_1=0} = g_1, \quad n = 0,1,2,\ldots. \tag{2.77}$$

Expressions analogous to Eqs. (2.76) and (2.77) are possible to write at other coordinate directions $x_2, x_3$. In particular, for the second coordinate direction $x_2$ it is necessary in Eqs. (2.76) and (2.77) to replace top and bottom indexes (1) and 1 on (2) and 2, and the top index (0) on (1). Thus for the initial condition in the second coordinate direction $x_2$ we have $v_H^{(2)}(0,x) = v^{(1)}(t_*, x)$.

Each function $v_{n+1}(t,x)$ in Eqs. (2.74) and (2.76) in the quasi-one-dimensional case or $v^{(1)}$ in Eqs. (2.76) and (2.77) there is a solution of linear equation that is a rather important feature of this algorithm. The algorithm comes from a method of Newton-Kantorovich's approximation [8] in functional space.

Corresponding restrictions on sizes $\frac{\partial f}{\partial v}, \frac{\partial f}{\partial \dot{v}}$ and others will be seen later. To reduce the further records we will introduce notations:

$$\begin{aligned}
&f_j = Y_j - \dot{v}^{(j)} a_j, \quad Y_1 = \sigma_1 z_1, \quad Y_2 = F_2 + \sigma_2 z_2,\\
&Y_3 = F_3 + \sigma_3 z_3, \quad c = v(m+1) + \alpha^{m+1}(x), \quad \omega = c^{\gamma},\\
&\frac{\partial u}{\partial v} = \frac{\omega}{c}, \quad \frac{\partial f_j}{\partial \dot{v}} = -a_j, \quad \phi_j = \frac{\partial f_j}{\partial v}, \quad j = 1,2,3,\\
&F_2 = a_5\omega^k, \quad F_3 = a_6 \exp(A_7\omega), \quad \phi_1 = 0,\\
&W_1 = \sigma_1 z_1, \quad \phi_2 = kF_2/c, \quad W_2 = F_2(1 - kv/c) + \sigma_2 z_2,\\
&\phi_3 = F_3\omega A_7/c, \quad W_3 = F_3(1 - A_7\omega v/c) + \sigma_3 z_3.
\end{aligned} \tag{2.78}$$

Let's find the solution of a linear boundary problem (2.76) and (2.77) at first on coordinate direction $x_1$, using the Eqs. (2.68), (2.69), and (2.78) with an index (1) above and below. We will substitute them in the first equation (2.76) and for simplicity of further calculations we will introduce the right-hand side of the received equation through $h_1 = W_1(v_n) - a_1\dot{v}^{(1)}$. Then it will become [36]:

$$\frac{\partial v^{(1)}}{\partial x_1} - \phi_1 v^{(1)} = h_1(v^{(0)}, \dot{v}^{(1)}, x, t). \tag{2.79}$$

If $\psi, g_j, j = 1,2,3$ from Eq. (2.51) or $E(u,x,t)$ in Eq. (2.42) are not introduced by spatial coordinates and time obviously instead of the algorithm (2.55)–(2.61) we will use the one-dimensional Laplace integral transformation [6].

Let's apply the Laplace integral transformation to the differential equation (2.79), excluding derivative on $x_1$ and replacing it with its linear expression concerning the image of required function for which Laplace integral transformation converges absolutely. The valid part of complex number $p = \xi + i\eta, i = \sqrt{-1}$, is considered positive, that is $\mathrm{Re}\, p > 0$. then using formulas from [6] and introducing images with symbols $V, H$, lowering an index (1) above and $\phi_1$ from Eq. (2.78), obviously independent on $x$ we have:

$$\begin{aligned}
&pV(t,p,x_2,x_3) - \phi_1 V(t,p,x_2,x_3) = g_1(t,x_2,x_3) + H_1(t,p,x_2,x_3),\\
&\quad 0 < x_j \le S_j, \quad j = 2,3
\end{aligned}$$

or

$$V = g_1/(p - \phi_1) + H_1/(p - \phi_1). \tag{2.80}$$

Using the Laplace integral transformation from [6]: $L^{-1}[(p - \phi_1)^{-1}] = \exp(\phi_1 x_1), L^{-1}[H_1(p)/p] = \int_0^{x_1} h_1(y)\, dy$, we restore the original for $v(x, t)$ from Eq. (2.80):

$$v^{(1)} = g_1 \exp(\phi_1 x_1) + \int_0^{x_1} \exp[\phi_1(x_1 - y)] h_1(v^{(0)}, \dot{v}^{(1)}, y, t)\, dy,$$
$$0 < x_j \leq S_j, \quad j = 2, 3. \tag{2.81}$$

Using the relationship again (2.78), the Eq. (2.81) will be rewritten in a detailed form:

$$\dot{v}^{(1)} + U_1 v^{(1)} = Z_1^{-1} \left\{ g_1 \exp(x_1 \phi_1) + \int_0^{x_1} \exp[\phi_1(x_1 - y)] \right.$$
$$\left. \times W_1(v^{(0)}, y, t)\, dy \right\} = R_1(v^{(0)}, x, t),$$

$$Z_1 = \int_0^{x_1} a_1(v^{(0)}) \exp[\phi_1(x_1 - y)]\, dy, \quad U_1 = U_1(v^{(0)}) = Z_1^{-1},$$
$$v^{(1)}(0, x) = v_H, \quad 0 < x_j \leq S_j, \quad j = 2, 3. \tag{2.82}$$

Finally the problem solution (2.82) [26] ($t = t_*$) will be:

$$v^{(1)}(t_*, x) = \left[ v_H^{(1)} + \int_0^{t_*} R_1(v^{(0)}, x, \tau) \exp(U_1 \tau)\, d\tau \right] \exp(-U_1 t_*),$$
$$v_H^{(1)} = v_H(x), \quad v^{(0)} = v_n, \quad n = 0, 1, 2, \ldots, \tag{2.83}$$

and $0 < x_j \leq S_j, \quad j = 2, 3$ are changed parametrically.

The same solutions have boundary problems (2.47) and (2.49) using Eqs. (2.74)–(2.77) and other Eqs. (2.70)–(2.73) and formulas (2.78). Then according to the algorithm (2.62)–(2.66) and (2.79)–(2.83) we have:

$$v^{(j)}(t_*, x) = \left[ v_H^{(j)} + \int_0^{t_*} R_j(v^{(j-1)}, x, \tau) \exp(U_j \tau)\, d\tau \right]$$
$$\times \exp(-U_j t_*), \quad v_H^{(j)} = v^{(j-1)}, \quad j = 2, 3, \tag{2.84}$$

where

$$U_j = Z_j^{-1}, \quad Z_j = \int_0^{x_j} a_j(v^{(j-1)}) \exp[\phi_j(x_j - y)]\, dy,$$
$$v^{(0)} = v_n, \quad n = 0, 1, 2, \ldots.$$

At $x = x_2$ in Eq. (2.84) other variables $0 < x_j \leq S_j, j = 1, 3$ are changed parametrically as in Eq. (2.83). A similar situation is for $x = x_3$; that is the final solution of a boundary problem (2.47) and (2.49): $v^{(3)}(x, t_*) = v_{n+1}(x, t_*) \forall t_* > 0, n = 0, 1, 2, \ldots$, and at presence of inversion formula (2.48) of initial nonlinear boundary problems (2.42)–(2.44).

## Existence, Uniqueness, and Convergence

We will consider the nonlinear case $m \geq 1, k > 1, w = u^m$ [36] and for simplicity of the analysis—range of definition $Q = \{x, t : 0 \leq x_j \leq b, b = \min(S_j), j = 1, 2, 3, 0 \leq t \leq t_k\}$ at $A_j = 1, j = \overline{1, 5}, A_6 = F = 0$. So the solution of the regional problem

$$\sum_{j=1}^{3} \frac{\partial uw}{\partial x_j} = u^k - \dot{u}, \quad u(0, x) = \alpha(x), \quad w = u^m;$$
$$u|_{x_1=0} = \beta_1, \quad u|_{x_2=0} = \beta_2, \quad u|_{x_3=0} = \beta_3 \tag{2.85}$$

using the algorithm (2.48), (2.68)–(2.84) for $g_j = 0$ is written on coordinate directions $x_j, j = 1, 2, 3$:

$$v^{(j)} = \left\{ v_H^{(j)} + \int_0^{t_*} [Z_j(v^{(j-1)})]^{-1} \left[ \int_0^{x_j} \exp[\phi_j(x_j - y)] \right.\right.$$
$$\left.\left. \times W_j(v^{(j-1)}, y, t_*)\, dy \right] \exp(U_j \tau) d\tau \right\} \exp(-U_j t_*),$$
$$j = 1, 2, 3, \quad v^{(0)} = v_n, \quad n = 0, 1, 2, \ldots, \tag{2.86}$$

where $v_H^{(1)} = v_H(x)$, $v_H^{(j)} = v^{(j-1)}$, $j = 2, 3$, $v^{(3)} = v_{n+1}(t_*, x)$;

$$U_j(v^{(j-1)}) = [Z_j(v^{(j-1)})]^{-1}, \quad Z_j = \int_0^{x_j} a_j(v^{(j-1)}) \exp[\phi_j(x_j - y)]\, dy,$$
$$j = 1, 2, 3, \quad W_j = \phi_j = 0, \quad j = 2, 3,$$

and $\phi_1 = \phi_1(F_2)$, $W_1 = W_1(F_2)$ are defined from Eq. (2.78). Using the algorithm (2.52)–(2.54), (2.62)–(2.66), the final solution from Eq. (2.86) will be (an index $*$ below at $t$ will be omitted):

$$v^{(3)}(t,x) = \exp[-tU(v_n)]\left\{v_H + \int_0^t [Z_1(v_n)]^{-1} \times \left[\int_0^{x_1} \exp[\phi_1(x_1-y)]W_1(v_n,y,t)\,dy\right]\exp[\tau U_1(v_n)]\,d\tau\right\},$$
$$U(v_n) = \sum_{j=1}^{3} U_j(v^{(j-1)}), \quad v^{(3)}(t,x) = v_{n+1}(t,x), \quad n = 0,1,2,\ldots. \tag{2.87}$$

It is supposed that in some neighborhood of a root function $f = f(v, \dot{v})$ from Eq. (2.74) together with the partial derivatives $\frac{\partial f}{\partial v}, \frac{\partial^2 f}{\partial v^2}, \frac{\partial f}{\partial \dot{v}}, \frac{\partial^2 f}{\partial \dot{v}^2}$ is continuous, and $\frac{\partial f}{\partial v}, \frac{\partial^2 f}{\partial v^2}, \frac{\partial f}{\partial \dot{v}}, \frac{\partial^2 f}{\partial \dot{v}^2}$ in this neighborhood does not go to zero.

Existence and uniqueness of solution (2.87) of a boundary problem (2.85) are proved in a similar manner to how it was made in Section 2.1 of this chapter [see formulas (2.19)–(2.30)].

## Estimation of Speed of Convergence [36]

We will consider at first the transfer equation from Eq. (2.85) with the help of Eq. (2.45) to the same type (2.47). Then we will apply locally-one-dimensional scheme splitting Eqs. (2.68)–(2.73) to the modified boundary problem (2.85) in the same manner as it was made for system (2.47) and (2.49), so we have with $\sigma = 1/3$:

$$\frac{\partial v^{(j)}}{\partial x_j} = -\frac{\gamma}{w^{(j-1)}}\frac{\partial v^{(j)}}{\partial t} + Y_j, \quad j = 1,2,3, \tag{2.88}$$
$$v^{(1)}(0,x) = v_H(x), \quad v^{(j)}(0,x) = v^{(j-1)}(t_*,x), \quad j = 2,3,$$
$$v^{(1)}|_{x_1=0} = 0, \quad v^{(2)}|_{x_2=0} = 0, \quad v^{(3)}|_{x_3=0} = 0,$$
$$Y_j = \sigma[\omega^{(j-1)}(v)]^k, \quad j = 1,2,3.$$

Let's address a recurrence relationship (2.74) and, noticing that $f(v,\dot{v}) = s(v) - r(\dot{v})$, we will subtract $n$-e the equation from $(n+1)$th to which, in a quasi-one-dimensional variant, corresponds to the first equation (2.76) for $v^{(1)}$, so we will receive on coordinate direction $x_1$:

$$\frac{\partial(v^{(1)} - v_n)}{\partial x_1} = s(v_n) - s(v_{n-1}) - (v_n - v_{n-1})\frac{\partial s(v_{n-1})}{\partial v} + (v^{(1)} - v_n)\frac{\partial s(v_n)}{\partial v} - \left[r(\dot{v}_n) - r(\dot{v}_{n-1}) - (\dot{v}_n - \dot{v}_{n-1})\frac{\partial r(\dot{v}_{n-1})}{\partial \dot{v}} + (\dot{v}^{(1)} - \dot{v}_n)\frac{\partial r(\dot{v}_n)}{\partial \dot{v}}\right]. \tag{2.89}$$

From the average theorem [33] it follows:

$$s(v_n) - s(v_{n-1}) - (v_n - v_{n-1})\frac{\partial s(v_{n-1})}{\partial v} = 0.5(v_n - v_{n-1})^2\frac{\partial^2 s(\xi)}{\partial v^2},$$
$$v_{n-1} \le \xi \le v_n.$$

Let's consider Eq. (2.89) how the equation is relative to $u^{(1)} = v^{(1)} - v_n(u^{(0)} = u_n, u_n = v_n - v_{n-1})$ and transform it as it was made above Eqs. (2.79)–(2.84). Then at $\phi_1 = \frac{\partial s(v_n)}{\partial v}, g_j = 0, j = 1, 2, 3$ we will have:

$$u^{(1)} = \int_0^{x_1} \exp[\phi_1(x_1 - y)]\left\{0.5\left[u_n^2\frac{\partial^2 s(v_n)}{\partial v^2} - \dot{u}_n^2\frac{\partial^2 r(\dot{v}_n)}{\partial \dot{v}^2}\right]\right.$$
$$\left. - \dot{u}^{(1)}\frac{\partial r(\dot{v}_n)}{\partial \dot{v}}\right\} dy$$

or

$$\dot{u}^{(1)} + U_1 u^{(1)} = 0.5 Z_1^{-1}\int_0^{x_1} \exp[\phi_1(x_1 - y)]\left[u_n^2\right.$$
$$\left.\times\frac{\partial^2 s(v_n)}{\partial v^2} - \dot{u}_n^2\frac{\partial^2 r(\dot{v}_n)}{\partial \dot{v}^2}\right] dy, \quad U_1 = Z_1^{-1}, \quad \dot{u}_n = \dot{u}^{(0)},$$
$$Z_1 = \int_0^{x_1} \exp[\phi_1(x_1 - y)]\frac{\partial r(\dot{v}_n)}{\partial \dot{v}} dy, \quad u_H = 0. \tag{2.90}$$

As a result, the problem solution (2.90) on the first coordinate direction $x_1$ will look like Eq. (2.83), where $v_H = 0$. It is also possible to receive the solution of the type (2.84) on coordinate directions $x_2, x_3$ from Eq. (2.88). The final solution similar to Eqs. (2.86) and (2.87), using the algorithm (2.62)–(2.66), (2.68)–(2.84) will be rewritten as:

$$u^{(3)}(t_*, x) = \exp(-t_* U)\int_0^{t_*} R_1 \exp(\tau U_1)\, d\tau$$
$$+ \exp[-t_*(U_2 + U_3)]\int_0^{t_*} R_2 \exp(\tau U_2)\, d\tau$$
$$+ \exp(-t_* U_3)\int_0^{t_*} R_3 \exp(\tau U_3)\, d\tau, \tag{2.91}$$

where

$$Z_j = \int_0^{x_j} \exp[\phi_j(x_j - y)]\frac{\partial r(\dot{v}^{(j-1)})}{\partial \dot{v}} dy, \quad U = \sum_{j=1}^{3} U_j(v^{(j-1)}),$$

$$R_j = 0.5Z_j^{-1}\int_0^{x_j}\exp[\phi_j(x_j-y)]\left[(u^{(j-1)})^2\frac{\partial^2 s(v^{(j-1)})}{\partial v^2}-(\dot{u}^{(j-1)})^2\right.$$
$$\left.\times\frac{\partial^2 r(\dot{v}^{(j-1)})}{\partial \dot{v}^2}\right]dy,\quad j=1,2,3,\quad U_j=Z_j^{-1},$$
$$u^{(0)}=u_n,\quad n=0,1,2,\ldots.$$

Let's put $\max\limits_{v,\dot{v}\in R}\left(\left|\frac{\partial s(v^{(j)})}{\partial v}\right|,\left|\frac{\partial r(\dot{v}^{(j)})}{\partial \dot{v}}\right|\right)=c_1,\ \max\limits_{v,\dot{v}\in R}\left(\left|\frac{\partial^2 s(v^{(j)})}{\partial v^2}\right|,\left|\frac{\partial^2 r(\dot{v}^{(j)})}{\partial \dot{v}^2}\right|\right)=c_2, j=0,1,2$, assuming $c_j<\infty, j=1,2$. We will use the results of the article [32] so we have: $U_1=Z_1^{-1}$, $\max\limits_{0\le x_1\le b} Z_1=\exp(c_1 b)-1=\nu>0$, $\max\limits_{0\le t\le t_k}\exp(-\nu t)\le 1$, $\max\limits_{0\le \tau\le t}\exp[-U_1(t-\tau)]\le 1$, $\frac{\partial^2 r(\dot{v}^{(j)})}{\partial \dot{v}^2}=0, j=0,1,2.$

Using the assumption on equality of all directions in space $(U_j=\nu, R_j=Bu_n^2, B=c_2/(2c_1), j=1,2,3)$ and functions $u^{(0)}=u^{(j)}, j=1,2$ (for converging sequence $v_n$ all intermediate values $u^{(j)}, j=0,1,2$ are close to zero as they are in a convergence interval: $[v^{(0)}, v^{(3)}]$), we have from Eq. (2.91) at $u_{n+1}(t_*,x)=u^{(3)}(t_*,x)$, lowering an index $(*)$ at $t$:

$$|u_{n+1}|\le 3u_n^2 Bt. \tag{2.92}$$

Let's choose $u_0(x,t)$ so that $|u_0(x,t)|\le 1$ in the region Q. As a result from expression (2.92) we will receive at $n=0$, introducing $M_1=\max\limits_Q|u_1|$:

$$M_1\le 3Bt = Y. \tag{2.93}$$

Then the top border $M_1$ will not surpass unit, if there is inequality $Y\le 1$ in Eq. (2.93)

$$t\le\frac{1}{3B}. \tag{2.94}$$

Therefore choosing an interval $[0,t]$ small enough [to satisfy the condition (2.94)] we have $M_1\le 1$. Finally we receive $M_{n+1}\le Yu_n^2$ or

$$\max_{x,t\in Q}|v_{n+1}-v_n|\le Y\max_{x,t\in Q}|v_n-v_{n-1}|^2. \tag{2.95}$$

The relationship (2.95) shows that if convergence in general takes place, it is quadratic. Thus, with big enough $n$ each following step doubles a number of correct signs in the given approximation.

## Results of Test Checks

Accuracy of the received approximate analytical formulas (2.83) and (2.84) we will check practically while solving the nonlinear boundary problem for the equation in partial derivatives in regions $Q = \{x, t : (0 \leq x_j \leq b, j = 1, 2, 3, 0 \leq t \leq t_k)\}$:

$$A_2 \frac{\partial u}{\partial t} + A_1 \sum_{j=1}^{3} \frac{\partial u^{m+1}}{\partial x_j} = A_3 u^k + A_4 \exp(A_5 u) + E(x, t); \tag{2.96}$$

$$u|_{t=0} = \exp\left(\sum_{j=1}^{3} y_j\right), \quad y_j = x_j/b; \tag{2.97}$$

$$u|_{x_1=0} = \exp(t + y_2 + y_3), \quad u|_{x_2=0} = \exp(t + y_1 + y_3),$$
$$u|_{x_3=0} = \exp(t + y_1 + y_2). \tag{2.98}$$

If the exact solution (2.96)–(2.98) is taken in a form of $u = \exp(t + z)$, $z = \sum_{j=1}^{3} y_j$, a source $E$ in the Eq. (2.96) will be

$$E = \exp(t + z)\{A_2 + 3A_1 b^{-1}(m + 1)\exp[m(t + z)]\}$$
$$- A_3 \exp[k(t + z)] - A_4 \exp[A_5 \exp(t + z)].$$

The following basic values of the initial data are taken: $m = 1$, $A_1 = A_2 = 1$, $b = 1$, $t_k = 1$, $N = 11$, $M = 201$, $h = b/(N - 1)$, $\tau = t_k/(M - 1)$; $N, M, h, \tau$ is a number of checkouts and steps on space and on time at finding integrals in the Eqs. (2.82)–(2.84) by trapezoid formula and Simpson's formula [21]. The program is made on Fortran-90 language, calculation was made on Pentium 2 (130 MHz, compiler PS 5) with double accuracy. In Table 2.2 there are results of calculations of maximum relative error $\varepsilon = \frac{|u-\tilde{u}|}{u} 100\%$ ($u$ is exact, $\tilde{u}$ is approximate analytical solution) and numbers of iterations $J$ ($||U_n|| \leq \delta$, $\delta = 1\%$) at various values $A_p, p = 3, 4, 5$, and $k$ for the basic variant. Time of calculation of any variant is equal to 5 s. The boundary problem (2.96)–(2.98) is solved with method quasi-linearization with the help of substitutions (2.48) and formulas (2.83) and (2.84). The number of iterations $J$ was traced on relative change of error vector (in percentage):

$$||U_n|| = \max_{x,t \in Q} |(v_{n+1} - v_n)/v_{n+1}|.$$

**Table 2.2** A dependence of the maximum relative errors at the various values $A_p, p = 3, 4, 5$, and $\kappa$

| | Results of calculations | | | | | |
|---|---|---|---|---|---|---|
| **Variant number** | $\kappa$ | $A_3$ | $A_4$ | $A_5$ | $\varepsilon, \%$ | $J$ |
| 1 | 1 | 0 | 0 | 0 | 4.87 | 2 |
| 2 | 1 | −1 | 0 | 0 | 3.9 | 2 |
| 3 | 2 | −1 | 0 | 0 | 6.3 | 2 |
| 4 | 2 | 1 | 0 | 0 | 10.6 | 2 |
| 5 | 1 | 1 | 0 | 0 | 5.82 | 2 |
| 6 | 1 | 0 | 1 | −1 | 4.86 | 2 |
| 7 | 1 | 1 | 1 | −1 | 5.84 | 2 |

# CHAPTER 3

# Method of Solution of Nonlinear Boundary Problems

In this chapter we have the approximate analytical formulas in the solution of equations in partial derivatives of the second order (concerning spatial coordinates). For nonlinear one-dimensional and three-dimensional first boundary problems we have found the estimation of rate of convergence of the iterative process. On the basis of trial functions we have results of test checks of mathematical technology and comparison with a known numerical method.

## 3.1 METHOD OF SOLUTION OF NONLINEAR BOUNDARY PROBLEMS

### Statement of a Problem and a Method Algorithm

Let us attempt to find the equation solution in partial derivatives of parabolic type [5, 7], with sources

$$C(T)\frac{\partial T}{\partial t} = \frac{\partial}{\partial x}\left[A(T)\frac{\partial T}{\partial x}\right] + B(T)\frac{\partial T}{\partial x} + A_1 T^k + A_2 \exp(A_4 T) + A_3(x,t) \tag{3.1}$$

in the region $Q = (0 < x < a, 0 < a < \infty)$, $\overline{Q} = Q + \Gamma$, $\overline{Q_t} = \overline{Q} \times [0 < t \leq t_0]$ at $A_j = \text{const}, j = 1, 2, 4$ with initial condition

$$T\,|_{t=0} = T_H(x) \tag{3.2}$$

on border $\Gamma$ with boundary conditions of the second, the third types

$$\left[A\frac{\partial T}{\partial x} + G_1(t)T\right]_{x=0} = D_1(t), \quad \left[A\frac{\partial T}{\partial x} + G_2(t)T\right]_{x=a} = D_2(t,a). \tag{3.3}$$

At $G_1 = G_2 = 0$ we have Neumann condition, a boundary Dirichlet condition is performed below in expression (3.39).

*Analytical Solution Methods for Boundary Value Problems*
http://dx.doi.org/10.1016/B978-0-12-804289-2.00003-X

Let's assume everywhere:

1. A problem (3.1)–(3.3) has the unique solution $T(x,t)$, which is continuous in the closed region $\overline{Q_t}$ and has continuous derivatives $\frac{\partial T}{\partial t}$, $\frac{\partial T}{\partial x}$, $\frac{\partial^2 T}{\partial x^2}$.
2. The following conditions are provided: $A(T) \geq c_1 > 0, C(T) \geq c_2 > 0$, $c_1, c_2$ are constants; $T_H$ is a continuous function in $\overline{Q}$, $A, B, A_3$ is a continuous functions in the closed region $\overline{Q_t}$.
3. Generally coefficients $C(T), B(T)$ can be nonlinearly dependent on the problem solution $T$ [5], the type $A(T)$ is defined below in the formula (3.5), $G_j, D_j, j = 1,2$ are continuous functions on $\Gamma$ for $0< t \leq t_0$ having limited partial derivatives of the first order.

   Let's use Kirchhoff's transformation [5]

$$v = \int_0^T \frac{A(T)}{A_H} dT, \tag{3.4}$$

where $A_H$ is, for example, the heat conductivity factor at a temperature equal to zero.

Further, that using the inversion formula, $A(T)$ in Eq. (3.1) we take in a form

$$A(T) = NT^m, \quad m > 0,\ N > 0,\ N = \text{const.} \tag{3.5}$$

Then, taking into account the relationships [5]:

$$\frac{\partial A}{\partial x} = \frac{\partial A}{\partial T}\frac{\partial T}{\partial x}, \quad \frac{\partial v}{\partial t} = \frac{A}{A_H}\frac{\partial T}{\partial t}, \quad \frac{\partial v}{\partial x} = \frac{A}{A_H}\frac{\partial T}{\partial x}, \tag{3.6}$$

we receive a boundary problem from Eqs. (3.1))–(3.6)

$$\frac{\partial^2 v}{\partial x^2} = c(v)\frac{\partial v}{\partial t} - b(v)\frac{\partial v}{\partial x} - a_1(\phi v)^{k/s} - a_2 \exp[A_4(\phi v)^{1/s}] - a_3(x,t), \tag{3.7}$$

$$v \,|_{t=0} = v_H(x), \quad v_H = \frac{T_H^s}{\phi}, \quad \phi = \frac{sA_H}{N}, \quad s = m+1, \tag{3.8}$$

$$\left[\frac{\partial v}{\partial x} + \frac{G_1(v\phi)^{1/s}}{A_H}\right]_{x=0} = \frac{D_1}{A_H}, \quad \left[\frac{\partial v}{\partial x} + \frac{G_2(v\phi)^{1/s}}{A_H}\right]_{x=a} = \frac{D_2}{A_H}. \tag{3.9}$$

Thus $T$ is defined from Eq. (3.4) according to the inversion formula

$$v = \frac{T^s}{\phi}, \quad T = (v\phi)^{1/s}, \tag{3.10}$$

transfer coefficients in Eq. (3.7) are written: $c = C/A, b = B/A, a_j = A_j/A_H, j = 1, 2, 3, A_4, a_1, a_2$ are constants.

Let's note that variation ranges of independent variables and type of the boundary conditions do not change in relation to Kirhhoff's transformation (3.4), and within the inversion formula (3.10) the boundary conditions of the first-third type pass into Dirichlet, Neumann, and Newton's conditions.

As $B(T)$ in Eq. (3.1) does not depend on $x$ we take substitution: $v = w\exp(-xb/2)$ [37] in Eqs. (3.7)–(3.9) to exclude the first partial derivative on space $x$ in Eq. (3.7). Then a boundary problem (3.7)–(3.9) and the inversion formula (3.10) will be rewritten

$$\frac{\partial^2 w}{\partial x^2} = c(w)\frac{\partial w}{\partial t} - \exp(rx)\{a_1[\phi w\exp(-rx)]^{k/s} + a_2\exp[A_4(\phi w\exp(-rx))^{1/s}] + a_3\} + wr^2, \tag{3.11}$$

$$w\,|_{t=0} = w_H(x), \quad w_H = \frac{T_H^s\exp(rx)}{\phi}, \quad r = 0.5b, \tag{3.12}$$

$$\left[\frac{\partial w}{\partial x} - rw + \frac{G_1(w\phi)^{1/s}}{A_H}\right]_{x=0} = \frac{D_1}{A_H},$$
$$\left[\frac{\partial w}{\partial x} - rw + \frac{\exp(ra)G_2(w\phi\exp(-ra))^{1/s}}{A_H}\right]_{x=a} = \frac{D_2\exp(ra)}{A_H}, \tag{3.13}$$

$$w = \frac{T^s\exp(rx)}{\phi}, \quad T = [w\phi\exp(-rx)]^{1/s}. \tag{3.14}$$

Our purpose is to have a solution to a nonlinear boundary problem if it exists, as a limit of sequence of solutions of linear boundary problems. For this we use the results of works [7, 15, 16]. Let $w_0 = \text{const}$ be some initial approximation (for an initial approximation it is better to take $w_H$ from Eq. (3.12)). We will consider sequence $w_n(t, x)$, defined by a recurrence relationship [7] (the point above corresponds to a partial derivative on time):

$$\frac{\partial^2 w_{n+1}}{\partial x^2} = f + (w_{n+1} - w_n)\frac{\partial f}{\partial w_n} + (\dot{w}_{n+1} - \dot{w}_n)\frac{\partial f}{\partial \dot{w}_n}, \quad f = f(w_n, \dot{w}_n), \tag{3.15}$$

$$w_H = w_n(0, x), \quad \frac{\partial w_{n+1}}{\partial x}|_\Gamma = \left[\alpha + (w_{n+1} - w_n)\frac{\partial \alpha}{\partial w_n}\right]\Big|_\Gamma,$$

$$\alpha = rw_n + \frac{1}{A_H}\exp(rx)\{D_i - G_i[w_n\phi\exp(-xr)]^{1/s}\}, \quad n = 0, 1, 2, \ldots, \tag{3.16}$$

where $i = 1$ at $x = 0$, $i = 2$ at $x = a$, the type $f$ will be defined below at the concrete solution of a boundary problem (3.11)–(3.13).

Each function $w_{n+1}(t, x)$ in Eqs. (3.15) and (3.16) is a solution of the linear equation that is a rather important feature of this algorithm. The algorithm comes from an approximation of Newton-Kantorovich's method [8] in functional space.

To reduce further records we will introduce the notations:

$$f = c\dot{w}_n - \exp(rx)[a_1 Z_n^{k/s} + a_2\exp(A_4 Z_n^{1/s}) + a_3] + w_n r^2,$$

$$\Phi = \frac{a_1 k Z_n^{k/s} A_H}{Z_n N} + \frac{a_2 A_4 Z_n^{1/s}\exp(A_4 Z_n^{1/s}) A_H}{Z_n N} - r^2, \quad r = 0.5\frac{B}{A},$$

$$R = a_1 Z_n^{k/s}\left[\exp(rx) - \frac{w_n k A_H}{Z_n N}\right] + a_2\exp(A_4 Z_n^{1/s})\left[\exp(rx) - \frac{w_n A_4 A_H}{Z_n N}\right] + a_3\exp(rx), \quad g_i = \frac{G_i Z_n^{1/s}}{Z_n N} - r,$$

$$Z_n = w_n\exp(-rx)\phi,$$

$$q_i(x,t) = \frac{\exp(rx)}{A_H}[D_i(t) - Z_n^{1/s}G_i(t)] + \frac{w_n G_i Z_n^{1/s}}{Z_n N}, \quad c(w) = \frac{\partial f}{\partial \dot{w}_n},$$
$$\Phi(w_n) = -\partial f/\partial w_n, \tag{3.17}$$

where for $g_i$ and $q_i$ : $i = 1$ at $x = 0$, $i = 2$ at $x = a$.

Let's take the solution of a boundary problem (3.11)–(3.13), using for this purpose the Eqs. (3.14)–(3.17) so introducing them, we have

$$h = c\dot{w}_{n+1} - R(w_n), \tag{3.18}$$

$$\frac{\partial^2 w_{n+1}}{\partial x^2} + w_{n+1}\Phi = h(w_n, \dot{w}_{n+1}, x, t), \quad w_n|_{t=0} = w_H, \quad n = 0, 1, 2, \ldots, \tag{3.19}$$

$$\left[\frac{\partial w_{n+1}}{\partial x} + g_1(t)w_{n+1}\right]_{x=0} = q_1(t), \tag{3.20}$$

$$\left[\frac{\partial w_{n+1}}{\partial x} + g_2(t,a)w_{n+1}\right]_{x=a} = q_2(t,a). \tag{3.21}$$

Let's consider $u(y)$ as a material or complex-place function of valid variable, defining in the region $S(0 \le y < \infty)$ and Lebesgue summable [38] in any final interval $S_z$ $(0 \le y \le z)$

$$F(p, z) = \int_0^z \exp(-py)u(y)\,dy, \tag{3.22}$$

where $p = \xi + i\eta$ is a complex parameter ($i = \sqrt{-1}$). Let $V$ be a set of all functions $u(y)$, for each of which there is such a parameter $p$ for which the conditions satisfied:

1. The integral (3.22) is limited in a point $p$ in relation to a variable $z \ge 0$

$$|F(p, z)| < M(p)$$

   for any $z \ge 0$, $M$ is a positive constant which is independent on $z$.
2. In a point $p$ there is a limit

$$\lim_{z \to \infty} F(p, z) = F(p).$$

If 1 and 2 are simultaneously provided, we may say [6, 39] that integral

$$F(p) = L_p|u(y)| = \int_0^\infty \exp(-py)u(y)\,dy$$

converges boundedly in a point $p$. Thus, the set $V$ consists of functions, for which the last integral converges boundedly in a point $p$.

It is supposed that in calculation of images on coordinate $x$ we operate with the functions analytically continued on values $x > a$ by that law which they are defined in an interval $(0, a)$.

Let's apply the Laplace integral transformation [6] to the differential equation (3.19), excluding derivative on $x$ replacing it with its linear expression concerning the image desired function. There are other functions for which the Laplace integral transformation converges absolutely. The valid part of complex number $p$ is considered positive, that is $\mathrm{Re}\, p > 0$. We will introduce images with capital letters $W, H$ and for simplicity of calculations put an index $n$ below at $w$. We will have:

$$W(t, p) = \int_0^\infty \exp(-px)w(t, x)\,dx, \quad w(t, x) = L^{-1}[W(t, p)],$$
$$h(t, x) = L^{-1}[H(t, p)].$$

We assume that the desired solution $w(t, x)$, and its derivatives entering into the Eq. (3.19), satisfy the conditions of the integrated Laplace transformation on $x$, and growth degree on $x$ functions $w(t, x)$ its derivatives do not depend on $t$. Multiplying members of Eq. (3.19) on $\exp(-px)$ integrating on $x$ from 0 to $\infty$, we will have [6, 15], introducing $\frac{\partial g}{\partial x} = \frac{\partial w(t,0)}{\partial x}$, $g = w(t,0)$. Note that $\Phi$ does not depend on $x$ in Eq. (3.17) ($\Phi$ is always possible to set on the bottom iteration on $n$, knowing values $w_n$ in initial and subsequent moments of time from formulas (3.12) and (3.14))

$$p^2 W(t,p) - pg - \frac{\partial g}{\partial x} + \Phi W(t,p) = H(t,p) \quad \text{or}$$

$$W = \frac{pg}{p^2 + b^2} + \frac{b\frac{\partial g}{\partial x}}{b(p^2+b^2)} + \frac{bH}{b(p^2+b^2)}, \qquad b = \sqrt{\Phi}. \tag{3.23}$$

Using the return Laplace integral transformation [6]: $L^{-1}[p/(p^2+b^2)] = \cos(bx)$ at $b^2 > 0$, $L^{-1}[p/(p^2-b^2)] = \cosh(bx)$ at $b^2 < 0$; $L^{-1}[H(p)/p] = \int_0^x h(y)\,dy$, let's restore the original for $w(t,x)$ from Eq. (3.23) [6]

$$w(t,x) = gu_1(x) + u_2(x)\frac{\partial g}{\partial x} + \int_0^x u_2(x-y)h(y)\,dy, \tag{3.24}$$

where $u_1(x) = \cos(bx)$, $u_2(x) = b^{-1}\sin(bx)$, $u_2(x-y) = b^{-1} \times \sin[b(x-y)]$ at $b^2 = \Phi > 0$ from Eq. (3.23); $u_1(x) = \cosh(bx)$, $u_2(x) = b^{-1}\sinh(bx)$, $u_2(x-y) = b^{-1}\sinh[b(x-y)]$ at $b^2 < 0$.

The case $\Phi = 0$ does not interest us, since all factors $A_1, A_2, B$ are not equal to zero simultaneously. Otherwise the Eq. (3.7) or Eq. (3.1) will be the differential equation with constant coefficients for which the solution under boundary conditions of the first-third type $C = \text{const}$ is known [37, 40].

To find the unknown derivative $\partial g/\partial x$ and functions $g$ in Eq. (3.24) we differentiate the last on $x$ (it is supposed that there is a limited partial derivative on $x$ from $w(t,x)$). Then we will have at $b = \sqrt{\Phi}$ from Eq. (3.17) obviously independent on $x$

$$\frac{\partial w}{\partial x} = \xi g u_2(x) + u_1(x)\frac{\partial g}{\partial x} + \int_0^x u_1(x-y)h(y)\,dy, \quad \xi = \text{sign } b^2 = \mp b^2, \tag{3.25}$$

where the sign "minus" undertakes at $\Phi > 0$, "plus"—at $\Phi < 0$ from Eq. (3.23).

Finally a derivative $\partial g/\partial x$ and function $g$ we find, if we use the Eqs. (3.20), (3.21), (3.24), and (3.25)

$$\left.\frac{\partial w}{\partial x}\right|_{x=0} = q_1 - g_1 w|_{x=0} = \frac{\partial g}{\partial x}, \tag{3.26}$$

$$\left.\frac{\partial w}{\partial x}\right|_{x=a} = q_2 - g_2 w|_{x=a} = \xi g u_2(a) + u_1(a)\frac{\partial g}{\partial x} + \int_0^a u_1(a-y)h(y)\,dy, \tag{3.27}$$

$$w(a) = g u_1(a) + u_2(a)\frac{\partial g}{\partial x} + \int_0^a u_2(a-y)h(y)\,dy. \tag{3.28}$$

Having substituted $w(a)$ from Eq. (3.28) in the Eq. (3.27), we find

$$g g_1 + \frac{\partial g}{\partial x} = q_1, \tag{3.29}$$

$$g B_1 + \frac{\partial g}{\partial x} B_2 = B_3, \tag{3.30}$$

$$B_1 = g_2 u_1(a) + \xi u_2(a), \quad B_2 = u_1(a) + g_2 u_2(a),$$
$$B_3 = q_2 - \int_0^a [u_1(a-y) + g_2 u_2(a-y)]h(y)\,dy. \tag{3.31}$$

Further, solving the system of the Eqs. (3.29) and (3.30), we have

$$g = \frac{B_2 q_1 - B_3}{\Delta}, \quad \frac{\partial g}{\partial x} = \frac{B_3 g_1 - B_1 q_1}{\Delta}, \quad \Delta = B_2 g_1 - B_1. \tag{3.32}$$

Let's substitute $g$ and $\partial g/\partial x$ from Eq. (3.32) in the Eq. (3.24) taking into consideration (3.31), we will have:

$$w = K + M\left\{q_2 - \int_0^a [u_1(a-y) + g_2 u_2(a-y)]h(y)\,dy\right\}$$
$$+ \int_0^x u_2(x-y)h(y)\,dy, \tag{3.33}$$

$$K = \frac{q_1[B_2 u_1(x) - u_2(x)B_1]}{\Delta}, \quad M = \frac{g_1 u_2(x) - u_1(x)}{\Delta}. \tag{3.34}$$

We transform the expression on the right-hand side (3.33) to get rid of the second integral with a variable top limit. Then, entering Green's function $G(x, y)$ [7, 15, 16]

$$G(x,y) = \begin{cases} M[u_1(a-y) + g_2 u_2(a-y)] - u_2(x-y), & 0 \le y \le x, \\ M[u_1(a-y) + g_2 u_2(a-y)], & x \le y \le a, \end{cases} \tag{3.35}$$

the expression (3.33), using formulas (3.17), (3.18), (3.34), and (3.35) will be rewritten as:

$$\dot{w}_{n+1} + Uw_{n+1} = Y^{-1}\left[\psi + \int_0^a G(x,y)R(w_n)\,dy\right]$$
$$= S(w_n, x, t), \quad n = 0, 1, 2, \ldots, \tag{3.36}$$

$$U = Y^{-1}, \; Y = \int_0^a c(w_n)G(x,y)\,dy, \; \psi = K + q_2M, \; w(0,x) = w_H(x). \tag{3.37}$$

Finally the problem solution (3.36) and (3.37) will be [33]:

$$w_{n+1}(x,t) = \left[w_H + \int_0^t S(w_n, x, \tau)\exp(\tau U)\,d\tau\right]$$
$$\times \exp(-tU), \quad n = 0, 1, 2, \ldots, \tag{3.38}$$

the solution $T(x,t)$ of the initial boundary problem (3.1)–(3.3) will agree with inversion formula (3.14).

For simplicity of further analysis we will write solution working formulas (3.38) for boundary conditions of the first type of problem (3.1):

$$T(0,t) = U_1(t), \quad T(a,t) = U_2(a,t), \tag{3.39}$$

where $U_i, i = 1, 2$ is a continuous function of time on border $\Gamma$.

Using the Kirhgof's transformation (3.4) and inversion formulas (3.10) and (3.14), we find instead (3.39)

$$w_{n+1}(0,t) = g, \quad w_{n+1}(a,t) = q, \tag{3.40}$$

where $g = U_1^s/\phi$, $q = U_2^s\exp(ra)/\phi$, the initial condition (3.2) preserves its previous view (3.19)

$$w_n(0,x) = w_H, \quad w_H = \frac{T_H^s\exp(rx)}{\phi}. \tag{3.41}$$

Then formulas (3.35) for $G(x,y)$ (3.37) for $\psi$ of the first boundary problem (3.19), (3.40), and (3.41) will be rewritten as:

$$\psi = gu_1(x) + \frac{u_2(x)}{u_2(a)}[q - gu_1(a)], \tag{3.42}$$

$$G(x,y) = \begin{cases} \frac{u_2(x)u_2(a-y)}{u_2(a)} - u_2(x-y), & 0 \le y \le x, \\ \frac{u_2(x)u_2(a-y)}{u_2(a)}, & x \le y \le a, \end{cases} \tag{3.43}$$

and a view of solution (3.36) and (3.38), formulas for $U$ from Eq. (3.37) $T$ from Eq. (3.14) will stay the same. That is the genericity of developed mathematical technology.

## Existence, Uniqueness, and Convergence

Without loss of generality, we will consider the solution of the simplified boundary problem (3.1), (3.2), and (3.39) at $U_i = 0, i = 1, 2$.

We take a nonlinear case $m > 0$, $k > 1$ in the region $Q = (0 < x < a)$, $\overline{Q} = Q + \Gamma$, $\overline{Q_t} = \overline{Q} \times [0 < t \leq t_0]$ at $A_2 = 0, C = 1, A_1 = -1$, $A(T) = T^m$

$$\frac{\partial T}{\partial t} = \frac{\partial}{\partial x}\left[A(T)\frac{\partial T}{\partial x}\right] + B(T)\frac{\partial T}{\partial x} - T^k, \quad 0 < t < t_0,$$
$$T(0, x) = T_H(x), \quad T|_\Gamma = 0. \tag{3.44}$$

After application of Kirhgof's transformation (3.4) at $A_H = 1$ and formulas (3.6) and (3.14) we have the modified boundary problem

$$c(w)\frac{\partial w}{\partial t} = \frac{\partial^2 w}{\partial x^2} - \exp(rx)a_1[sw\exp(-rx)]^{k/s} - wr^2, \quad 0 < t < t_0,$$
$$w\,|_{t=0} = w_H, \quad w_H = \frac{T_H^s \exp(rx)}{s}, \quad w|_\Gamma = 0, \quad r = 0.5\frac{B}{A}. \tag{3.45}$$

Thus $w$ is defined from Eq. (3.14) according to the inversion formula:

$$w = \frac{T^s \exp(rx)}{s}, \quad T = [ws\exp(-rx)]^{1/s}. \tag{3.46}$$

As a result of algorithm application (3.15)–(3.19), (3.23), (3.24), (3.28), (3.36), (3.38), (3.41)–(3.43) to a boundary problem (3.45) we have

$$w_{n+1}(x, t) = \left\{w_H + \int_0^t S(w_n, x, \tau)\exp[\tau U(w_n)]\,d\tau\right\}$$
$$\times \exp[-tU(w_n)], \quad n = 0, 1, 2, \ldots, \tag{3.47}$$

$$S(w_n, x, t) = Y^{-1}\left[\psi + \int_0^a G(x, y)R(w_n)\,dy\right], \quad \Phi = -kZ_n^{k/s-1} - r^2,$$

$$R = -Z_n^{k/s}\left[\exp(rx) - \frac{w_n k}{Z_n}\right], \quad \psi = 0, \quad Z_n = sw_n\exp(-xr),$$

$$Y(w_n) = \int_0^a c(w_n)G(x, y)\,dy, \quad c = A^{-1}(w), \quad U = Y^{-1}, \tag{3.48}$$

where $G(x, y)$ preserves its view (3.43).

The final solution $T$ of a boundary problem (3.44) we receive after substitutions $w_{n+1}$ from Eq. (3.47) in the inversion formula (3.46).

**Theorem.** *Let $w$ be continuously differentiated in $\overline{Q_t}$, then in the region $\overline{Q_t}$ there is the unique solution of the modified boundary problem* (3.45).

*Existence and uniqueness of the solution of a boundary problem* (3.44) *are proved in the same manner as in Chapter 2 [see formulas* (2.19)–(2.30)*].*

## Estimation of Speed of Convergence [15, 16]

It is considered that in some neighborhood of a root function $f = f(w,\dot{w})$ from Eq. (3.15) together with the partial derivatives $\partial f/\partial w$, $\partial^2 f/\partial w^2, \partial f/\partial \dot{w}$, $\partial^2 f/\partial \dot{w}^2$ are continuous, $\partial f/\partial w$, $\partial^2 f/\partial w^2$, $\partial f/\partial \dot{w}$, $\partial^2 f/\partial \dot{w}^2$ in this neighborhood do not go to zero.

Let's address the recurrence relationship (3.15), noticing that $f(w,\dot{w}) = s(w) + r(\dot{w})$ in Eq. (3.15), we will subtract $n$-e the equation from $(n+1)$th then we will find:

$$\begin{aligned}\frac{\partial^2(w_{n+1}-w_n)}{\partial x^2} &= s(w_n) - s(w_{n-1}) - (w_n - w_{n-1})\frac{\partial s(w_{n-1})}{\partial w} \\ &+ (w_{n+1}-w_n)\frac{\partial s(w_n)}{\partial w} + [r(\dot{w}_n) - r(\dot{w}_{n-1}) \\ &- (\dot{w}_n - \dot{w}_{n-1})\frac{\partial r(\dot{w}_{n-1})}{\partial \dot{w}} + (\dot{w}_{n+1}-\dot{w}_n)\frac{\partial r(\dot{w}_n)}{\partial \dot{w}}\Bigg].\end{aligned} \tag{3.49}$$

From the average theorem [33] it follows:

$$\begin{aligned}s(w_n) - s(w_{n-1}) - (w_n - w_{n-1})\frac{\partial s(w_{n-1})}{\partial w} &= 0.5(w_n - w_{n-1})^2 \\ \times \frac{\partial^2 s(\xi)}{\partial w^2}, \quad w_{n-1} \le \xi \le w_n.&\end{aligned}$$

Let's consider Eq. (3.49) how the equation is in relation to $u_{n+1} = w_{n+1} - w_n$, and transform it as above Eqs. (3.45)–(3.48). Then we will have:

$$\begin{aligned}u_{n+1} = \int_0^a G(y,x)\Bigg\{&\left[u_n^2\frac{\partial^2 s(w_n)}{\partial w^2} + \dot{u}_n^2\frac{\partial^2 r(\dot{w}_n)}{\partial \dot{w}^2}\right]/2 \\ &+ u_{n+1}\frac{\partial s(w_n)}{\partial w} + \dot{u}_{n+1}\frac{\partial r(\dot{w}_n)}{\partial \dot{w}}\Bigg\}\,dy\end{aligned}$$

or

$$\begin{aligned}\dot{u}_{n+1} - u_{n+1}\left[1 - \int_0^a G(x,y)\frac{\partial s(w_n)}{\partial w}\,dy\right]Y^{-1} &= -0.5Y^{-1}\int_0^a G(x,y) \\ \times\left[u_n^2\frac{\partial^2 s(w_n)}{\partial w^2} + \dot{u}_n^2\frac{\partial^2 r(\dot{w}_n)}{\partial \dot{w}^2}\right]dy, \quad u_H = 0,&\end{aligned}$$

$$Y = \int_0^a G(x,y)\frac{\partial r(\dot{w}_n)}{\partial \dot{w}}\,dy. \tag{3.50}$$

$$G(x,y) = \begin{cases} \frac{y(x-a)}{a}, & 0 \le y \le x, \\ \frac{x(y-a)}{a}, & x \le y \le a. \end{cases}$$

Finally the problem solution (3.50) will look like Eq. (3.47), where $w_H = 0$.

Let's put $\max\limits_{w,\dot{w}\in V}\left(\left|\frac{\partial s(w)}{\partial w}\right|, \left|\frac{\partial r(\dot{w})}{\partial \dot{w}}\right|\right) = c_1$, $\max\limits_{w,\dot{w}\in V}\left(\left|\frac{\partial^2 s(w)}{\partial w^2}\right|, \left|\frac{\partial^2 r(\dot{w})}{\partial \dot{w}^2}\right|\right) = c_2$, $\max\limits_{x,y}|G(x,y)| = a/4$ [7] assuming $c_m < \infty$, $m = 1.2$. Then from the Eqs. (3.47) and (3.50) it follows that $\frac{\partial^2 r(\dot{w})}{\partial \dot{w}^2} = 0$:

$$\|u_{n+1}| \le B\exp(t\alpha)\int_0^t u_n^2 \exp(-\alpha\tau)\,d\tau. \tag{3.51}$$

Let's choose $u_0(t,x)$ so that $|u_0(t,x)| \le 1$ is in the region $\overline{Q_t}$. As a result from expression (3.51) at $n = 0$, introducing $M_1 = \max\limits_{\overline{Q_t}}|u_1|$, we find: $z_0 = \max\limits_{\overline{Q_t}} u_0^2$, $z_0 \le 1, B = c_2/2c_1, Y = c_1 a^2/4, \alpha = 4/c_1 a^2 - 1$:

$$M_1 \le \frac{B[\exp(t\alpha) - 1]}{\alpha} = S. \tag{3.52}$$

Hence, under a condition $\alpha > 0$ ($a < 2/\sqrt{c_1}$) we find, that the top border $M_1$ will not surpass 1, if there is inequality $S \le 1$ in Eq. (3.52):

$$t \le \ln\left(\frac{\alpha}{B} + 1\right)^{1/\alpha}. \tag{3.53}$$

Therefore if choosing an interval $[0,t]$, $[0,a]$ small enough so that it satisfies the condition (3.53), we will have $M_1 \le 1$. Finally we receive definitively $M_{n+1} \le Sz_n$ or

$$\max_{x,t\in\overline{Q_t}}|w_{n+1} - w_n| \le S\max_{x,t\in\overline{Q_t}}|w_n - w_{n-1}|^2. \tag{3.54}$$

The relationship (3.54) shows that if convergence of the iterative process for a boundary problem (3.45) or (3.44) according to the inversion formula (3.46) in general takes place, it is quadratic. Thus, with a big enough $n$ each following step doubles a number of correct signs in the given approximation.

## Results of Test Checks

Estimation of error of analytical formulas (3.14), (3.36)–(3.38), (3.42), and (3.43) are checked practically in the solution of a boundary problem for the equation in partial derivatives in the region $\overline{Q_t} : [0 \le x \le a, 0 \le t \le t_0]$

$$A_6 \frac{\partial T}{\partial t} = \frac{\partial}{\partial x}\left[A_2(T)\frac{\partial T}{\partial x}\right] + A_1 \frac{\partial T}{\partial x} + A_3 T^k + A_4 \exp(A_5 T) + F(x, t), \tag{3.55}$$

$$T|_{t=0} = \exp(z), \quad z = \frac{x}{a}, \tag{3.56}$$

$$T|_{x=0} = \exp(\tau), \quad T|_{x=a} = \exp(\tau + 1), \quad \tau = \frac{t}{c}. \tag{3.57}$$

We have taken the exact solution of a problem (3.55)–(3.57)

$$T = \exp(\tau + z), \tag{3.58}$$

then the source $F$ in the Eq. (3.55) will be

$$F = \exp(\tau + z)\left\{\frac{A_6}{c} - \frac{A_1}{a} - \frac{(m+1)\exp[m(\tau + z)]}{a^2}\right\}$$
$$- A_3 \exp[k(\tau + z)] - A_4 \exp[A_5 \exp(\tau + z)], \quad A_2 = T^m.$$

The following basic values of initial data were used: $m = 0.5$, $t_0 = 1$, $A_H = 1, N = 11, \Delta x = a/(N-1), \Delta t = 0.5$ are numbers of checkouts steps on space and time at finding integrals in the Eqs. (3.36)–(3.38) by Simpson's formula [21].

The boundary problem (3.55)–(3.57) is solved by means of formulas (3.14), (3.36)–(3.38), (3.42), and (3.43). The number of iterations was traced [for total expressions of a type (3.38) and (3.47)] on relative change of an error vector:

$$||V_n|| = \max_{x,t \in \overline{Q_t}} \left|\frac{w_{n+1} - w_n}{w_{n+1}}\right|.$$

The program is made in the language Fortran-90, calculation was made on Pentium 4 (3.5 GHz, compiler PS 5) with double accuracy. Table 3.1 gives the maximum relative error

$$\varepsilon = \frac{|T - \tilde{T}|100\%}{T}, \tag{3.59}$$

**Table 3.1** A dependence of the maximum relative errors at the various values $A_j, j = 1, 3, 4, 5, k$

| Variant number | Results of calculations | | | | | | |
|---|---|---|---|---|---|---|---|
| | $k$ | $A_1$ | $A_3$ | $A_4$ | $A_5$ | $\varepsilon_1$, % | $\varepsilon_2$, % |
| 1 | 1 | 0.1 | 0 | 0 | 0 | 9.27 | 5.14 |
| 2 | 1 | −0.1 | 0 | 0 | 0 | 13.5 | 5.5 |
| 3 | 1 | 0.1 | 1 | 0 | 0 | 9.94 | 5.14 |
| 4 | 2 | 0.1 | 1 | 0 | 0 | 13.4 | 5.14 |
| 5 | 1 | −0.1 | 1 | 0 | 0 | 8.89 | 5.14 |
| 6 | 2 | −0.1 | 1 | 0 | 1 | 16.5 | 5.5 |
| 7 | 1 | −0.1 | −1 | 0 | 0 | 12.9 | 5.5 |
| 8 | 1 | 0.1 | 0 | 1 | −1 | 8.95 | 5.14 |
| 9 | 1 | 0.1 | 1 | 1 | −1 | 9.38 | 5.14 |

where $T$ is the exact obvious solution (3.58), $\tilde{T}$ is the approximate analytical solution on mathematical technology at various values $k$, $A_j, j = 1, 3, 4, 5$. In Eq. (3.59) $\varepsilon_1$ answers $a = 1$, $c = 5$, $A_6 = 10$, $\varepsilon_2$—$a = 0.01$, $c = 1$, $A_6 = 9 \cdot 10^3$ (a material of type of copper [31]) to the same basic data.

In Table 3.1 there are the results of test calculations $||V_n|| \leq \delta$, $\delta = 0.01$. Thus only two-three iterations for achievement of this accuracy were required; time of calculation of any variant is $t_p = 1$ s.

As is seen from Table 3.1 calculation on developed mathematical technology has almost the small error $\varepsilon_2$ from Eq. (3.59).

In comparison with the numerical solution of a problem (3.55)–(3.57) the scheme of absolutely steady difference with error of approximation for the first-second derivative on space—$O[(\Delta x)^2]$ [12] the two-layer scheme for derivative on time with a margin error approximations—$O(\Delta t)$ was used.

For $\varepsilon_1(a = 1, A_6 = 10)$ at $N = 41, \Delta t = 0.5$ and other identical initial data from Table 3.1 at number 4, numerical solution gives $\varepsilon_1 = 20.3\%$ (by the time of time $t_0 = 1$), $t_p = 2$ s. On mathematical technology from this section for the same initial sizes from Table 3.1 at number 4 we have 13.4%.

For $\varepsilon_2(a = 0.01, A_6 = 9 \cdot 10^3)$ under the numerical solution we receive 9.4%, on mathematical technology section from Table 3.1 at number 4 we have 5.14%. At the numerical solution of a problem (3.55) and (3.57) to receive accuracy of an order $\varepsilon_2 = 5\%$, it is necessary to take a step on time $\Delta t$ twice less ($\Delta t = 0.25$). Thus calculation of time $t_p = 4$ s also increases twice. This time $t_p$ essentially (on an order) increases in numerical solution of three-dimensional boundary problems [22].

## 3.2 METHOD OF SOLUTION OF THREE-DIMENSIONAL NONLINEAR FIRST BOUNDARY PROBLEM

### Statement of a Problem and a Method Algorithm

Let us attempt to find the solution of the equation of a parabolic type [2, 41] with nonlinear sources [2]

$$C(T)\frac{\partial T}{\partial t} = \operatorname{div}[A(T)\nabla T] + B(T)\nabla T + A_1 T^k + A_2\exp(A_4 T) + A_3(x,t) \tag{3.60}$$

in the region $Q = (0 < x_j < L_j; 0 < L_j < \infty, j = 1,2,3)$, $\overline{Q} = Q + \Gamma$, $\overline{Q_t} = \overline{Q} \times [0 < t \le t_k]$, $\Gamma$ is a boundary surface of ranges of definition Q; $A_j = \text{const}, j = 1,2,4$ (for simplicity of the analysis in the absence of mixed derivatives) with the initial condition

$$T|_{t=0} = T_H(x) \tag{3.61}$$

and with a boundary condition of the 1st type

$$T|_{\Gamma} = \Psi, \tag{3.62}$$

where $\Psi \neq \text{const}$. The equations of a type (3.60) are applied in heat conductivity theory [5] and in the mechanics of reacting [2] environments.

Let's assume everywhere:

1. A problem (3.60)–(3.62) has the unique solution $T(x,t)$, which is continuous in the closed region $\overline{Q_t}$ also has continuous derivatives $\frac{\partial T}{\partial t}$, $\frac{\partial T}{\partial x_j}, \frac{\partial^2 T}{\partial x_j^2}, j = 1,2,3$.
2. The following conditions are satisfied: $A(T) \ge c_1 > 0$, $C(T) \ge c_2 > 0$, $c_1, c_2$ are constants, $T_H$ is a continuous function in $\overline{Q}$, and $A, B, A_3$ is a continuous function in the closed region $\overline{Q_t}$.
3. $\Psi$ is a continuous function on sides $\Gamma$ for $0 \le t \le t_k$, having the limited partial derivatives of the first order.

Further, to use the inversion formula, we take

$$A(T) = s + zT^m, \tag{3.63}$$

where ($m > -1$, $s > 0$, $z$ are constants [5]).

Let's apply the Kirchhoff's transformation [5]

$$v = \int_0^T \frac{A(T)}{A_H}\, dT, \tag{3.64}$$

where $A_H$ is for example, the heat conductivity factor at the temperature equal to zero. Then, taking into account the relation [5]

$$\nabla A = \frac{\partial A}{\partial T}\nabla T, \quad \frac{\partial v}{\partial t} = \frac{A}{A_H}\frac{\partial T}{\partial t}, \quad \nabla v = \frac{A}{A_H}\nabla T, \tag{3.65}$$

we receive the differential equation from Eqs. (3.60) and (3.65):

$$\nabla^2 v = c(v)\partial v/\partial t - b(v)\nabla v - a_1 T^k - a_2 \exp(A_4 T) - a_3(x, t), \tag{3.66}$$

where $c = C/A, b = B/A, a_j = A_j/A_H, j = 1, 2, 3, A_4, a_1, a_2$ are constants.

Then we have from Eqs. (3.63) and (3.64) if using Newton's method [7]:

$$\begin{aligned}
&A_H(m+1)v = s(m+1)T + zT^{m+1},\\
&\quad T_{n+1} = T_n - \frac{f(T_n)}{f'_T(T_n)}, \quad f'_T(T_n) = \frac{\partial f(T_n)}{\partial T_n},\\
&\quad f(T_n) = zT_n^{m+1} + s(m+1)T_n - A_H(m+1)v(t, x),\\
&\quad f'_T(T_n) = z(m+1)T_n^m + s(m+1), \quad n = 0, 1, 2, \ldots.
\end{aligned} \tag{3.67}$$

Here as initial approximation $T_0 = \text{const}$ is any constant number close to $T_H$ from Eq. (3.61), and $v$ is defined below [see (3.103)]. Using formulas (3.64) and (3.65) initial and boundary conditions (3.61) and (3.62) for the Eq. (3.66) will be rewritten as:

$$v|_{t=0} = v_H, \quad v_H = [sT_H + zT_H^{m+1}/(m+1)]/A_H = F(T_H); \tag{3.68}$$

$$v|_\Gamma = A_H^{-1}[sT + zT^{m+1}/(m+1)]|_\Gamma = \Phi. \tag{3.69}$$

For the idea of a method of reduction of a multidimensional problem to a sequence of one-dimensional problems [12], we will show an example of the solution of a linear boundary problem for a three-dimensional equation of heat conductivity in the absence of a source:

$$\frac{\partial T}{\partial t} = \alpha \sum_{j=1}^{3} \frac{\partial^2 T}{\partial x_j^2}, \quad 0 < x_j < L_j, \quad j = 1, 2, 3, \quad t > 0; \tag{3.70}$$

$$T|_\Gamma = 0, \quad T|_{t=0} = T_H(x), \quad [\alpha] = \text{m}^2/\text{s}, \tag{3.71}$$

where $\Gamma$ is a border of a parallelepiped in which the boundary problem is defined (3.70) and (3.71).

Let's consider locally-one-dimensional scheme splittings [12]:

$$R_{(j)} V_{(j)} = \alpha \frac{\partial^2 T}{\partial x_j^2}; \tag{3.72}$$

$$\frac{\partial V_{(1)}}{\partial t} = R_{(1)} V_{(1)}, \quad 0 < t < t_*, \quad V_{(1)}(0, x) = T_H(x), \quad V_{(1)}|_\Gamma = 0; \tag{3.73}$$

$$\frac{\partial V_{(2)}}{\partial t} = R_{(2)} V_{(2)}, \quad 0 < t < t_*, \quad V_{(2)}(0, x) = V_{(1)}(t_*, x), \quad V_{(2)}|_\Gamma = 0; \tag{3.74}$$

$$\frac{\partial V_{(3)}}{\partial t} = R_{(3)} V_{(3)}, \quad 0 < t < t_*, \quad V_{(3)}(0, x) = V_{(2)}(t_*, x), \quad V_{(3)}|_\Gamma = 0. \tag{3.75}$$

The analytical solution of problems (3.72)–(3.75) by the Fourier method is presented by formulas [18, 40]

$$V_{(1)}(x, t_*) = \frac{2}{L_1} \sum_{n_1=1}^{\infty} \int_0^{L_1} T_H(\xi, x_2, x_3) \sin\left(\frac{n_1 \pi \xi}{L_1}\right) d\xi \\ \times \exp(-t_* Y_1) \sin(X_1); \tag{3.76}$$

$$V_{(2)}(x, t_*) = \frac{2}{L_2} \sum_{n_2=1}^{\infty} \int_0^{L_2} V_{(1)}(t_*, x_1, \eta, x_3) \sin\left(\frac{n_2 \pi \eta}{L_2}\right) d\eta \\ \times \exp(-t_* Y_2) \sin(X_2); \tag{3.77}$$

$$V_{(3)}(x, t_*) = \frac{2}{L_3} \sum_{n_3=1}^{\infty} \int_0^{L_3} V_{(2)}(t_*, x_1, x_2, \zeta) \sin\left(\frac{n_3 \pi \zeta}{L_3}\right) d\zeta \\ \times \exp(-t_* Y_3) \sin(X_3), \tag{3.78}$$

where $X_j = n_j x_j \pi / L_j$, $Y_j = n_j^2 \pi^2 \alpha / L_j^2$, $j = 1, 2, 3$.

Let's substitute $V_{(1)}$ from Eq. (3.76) in Eq. (3.77), and then $V_{(2)}$ from Eq. (3.79) in the Eq. (3.78) then we will have:

$$V_{(2)}(x, t_*) = \frac{4}{L_1 L_2} \sum_{n_2=1}^{\infty} \sum_{n_1=1}^{\infty} \exp[-t_*(Y_1 + Y_2)] \int_0^{L_2} \sin\left(\frac{n_2 \pi \eta}{L_2}\right) \\ \times \left[\int_0^{L_1} T_H(\xi, \eta, x_3) \sin\left(\frac{n_1 \pi \xi}{L_1}\right) d\xi\right] d\eta \sin(X_1) \sin(X_2); \tag{3.79}$$

$$V_{(3)}(x, t_*) = \frac{8}{L_1 L_2 L_3} \sum_{n_3=1}^{\infty} \sum_{n_2=1}^{\infty} \sum_{n_1=1}^{\infty} \int_0^{L_3} \sin\left(\frac{n_3 \pi \zeta}{L_3}\right) \left\{ \int_0^{L_2} \sin\left(\frac{n_2 \pi \eta}{L_2}\right) \right.$$

$$\times \left[ \int_0^{L_1} T_H(\xi, \eta, \zeta) \sin\left(\frac{n_1 \pi \xi}{L_1}\right) d\xi \right] d\eta \Big\} d\zeta \exp\left(-t_* \sum_{j=1}^{3} Y_j\right)$$

$$\times \prod_{j=1}^{3} \sin(X_j) = T(t_*, x), \quad \forall t_* > 0. \tag{3.80}$$

$T(t_*, x)$ in expression (3.80), as it is known [42], there is a solution of a boundary problem (3.70) and (3.71) under certain conditions of smoothness of initial and boundary conditions. With this model, the process of heat conductivity "is stretched out" in time and occurs at a time interval $3t_*$, instead of $t_*$ [13].

Let's apply locally-one-dimensional scheme splitting to Eq. (3.66) at differential level [12] and for simplicity of the analysis we will consider the first boundary problem:

$$\sum_{j=1}^{3} \frac{\partial^2 v}{\partial x_j^2} = c\frac{\partial v}{\partial t} - b\sum_{j=1}^{3} \frac{\partial v}{\partial x_j} - a_1 T^k - a_2 \exp(A_4 T) - a_3(x, t), \tag{3.81}$$

$$v_H = F(T_H), \quad v|_\Gamma = \Phi. \tag{3.82}$$

As a result we have:

$$\frac{\partial^2 v^{(1)}}{\partial x_1^2} = c\frac{\partial v^{(1)}}{\partial t} - b\frac{\partial v^{(1)}}{\partial x_1} - \sigma_1 a_3, \quad 0 < t < t_*; \tag{3.83}$$

$$v^{(1)}(0, x) = v_H(x), \quad v^{(1)}|_{x_1=0} = g_1(t, x_2, x_3),$$

$$v^{(1)}|_{x_1=L_1} = q_1(t, L_1, x_2, x_3); \tag{3.84}$$

$$\frac{\partial^2 v^{(2)}}{\partial x_2^2} = c\frac{\partial v^{(2)}}{\partial t} - b\frac{\partial v^{(2)}}{\partial x_2} - a_1 (T^{(1)})^k - \sigma_2 a_3, \quad 0 < t < t_*; \tag{3.85}$$

$$v^{(2)}(0, x) = v^{(1)}(t_*, x), \quad v^{(2)}|_{x_2=0} = g_2(t, x_1, x_3),$$

$$v^{(2)}|_{x_2=L_2} = q_2(t, x_1, L_2, x_3); \tag{3.86}$$

$$\frac{\partial^2 v^{(3)}}{\partial x_3^2} = c\frac{\partial v^{(3)}}{\partial t} - b\frac{\partial v^{(3)}}{\partial x_3} - a_2 \exp(A_4 T^{(2)})$$

$$- \sigma_3 a_3, \quad 0 < t < t_*; \tag{3.87}$$

$$v^{(3)}(0, x) = v^{(2)}(t_*, x), \quad v^{(3)}|_{x_3=0} = g_3(t, x_1, x_2),$$

$$v^{(3)}|_{x_3=L_3} = q_3(t, x_1, x_2, L_3), \tag{3.88}$$

where $\sigma_1 + \sigma_2 + \sigma_3 = 1$, and $T(v)$ is defined from Eq. (3.67).

Our purpose is to find the solution of a nonlinear boundary problem, if it exists, as a limit of sequence of solutions of the linear boundary problems. For this purpose we will use the results of work [7]. Let's note that in the article [10] usage of method quasi-linearization (Newton-Kantorovich's method) is based on a formal calculation of Frechet differential [43] that leads to more difficult relationships in comparison with [7]. Besides, in [10] the algorithm of solution of three-dimensional nonlinear boundary problem is absent. Let's assume further that all coordinate directions in space are equivalent.

Let $v_0 = \text{const}$ be an initial approximation [as an initial approximation it is better to take the value close to $v_H$ from Eq. (3.68)]. We will consider for simplicity of the analysis the quasi-one-dimensional case and sequence $v_n(t,x)$, defining recurrence relationship [7] (a point above and a stroke on the right correspond to a partial derivative on time and on space accordingly):

$$\frac{\partial^2 v_{n+1}}{\partial y^2} = f + (v_{n+1} - v_n)\frac{\partial f}{\partial v} + (v'_{n+1} - v'_n) \times \frac{\partial f}{\partial v'} + (\dot{v}_{n+1} - \dot{v}_n)\frac{\partial f}{\partial \dot{v}}, \quad f = f(v_n, \dot{v}_n, v'_n); \tag{3.89}$$

$$v_H = v_n(0,x), \quad v_{n+1}|_\Gamma = \Phi, \quad n = 0,1,2,\ldots, \tag{3.90}$$

where $y$ is any of coordinates $x_j, j = 1,2,3$ in Eq. (3.89). Then at $y = x_1$ other coordinates in Eq. (3.89), $0 < x_j < L_j, j = 2,3$ are changed parametrically. On the remained coordinates while receiving expressions (3.89) there is a circular replacement of indexes, if instead of $y$ to substitute accordingly $x_2, x_3$. We will note that in the solution of the three-dimensional boundary problem (3.81) and (3.82), if in the first coordinate direction $x_1$ the initial iterations acts as $v_n$, according to formulas (3.76)–(3.80) the subsequent iteration $v_{n+1}$ is from the final expression of type (3.80) in which it is necessary to put $v_{n+1}(t_*, x) = v^{(3)}(t_*, x)$ [see comments on the formula (3.103)]. Then in the quasi-one-dimensional equation variant (3.89) and (3.90) coordinate $x_1$ will be rewritten as [44]:

$$\frac{\partial^2 v^{(1)}}{\partial x_1^2} = f_1 + (v^{(1)} - v^{(0)})\frac{\partial f_1}{\partial v} + (v'^{(1)} - v'^{(0)})\frac{\partial f_1}{\partial v'} + (\dot{v}^{(1)} - \dot{v}^{(0)})\frac{\partial f_1}{\partial \dot{v}}, \quad v^{(0)} = v_n, \quad f_1 = f_1(v^{(0)}, \dot{v}^{(0)}, v'^{(0)}); \tag{3.91}$$

$$v_H^{(1)} = v_H(x), \quad v^{(1)}|_{x_1=0} = g_1, \quad v^{(1)}|_{x_1=L_1} = q_1, \quad n = 0, 1, 2, \ldots . \tag{3.92}$$

Formulas (3.91) and (3.92) are possible to write at other coordinate directions $x_2, x_3$. In particular, for the second coordinate direction $x_2$ it is necessary in Eqs. (3.91) and (3.92) to replace top and bottom indexes (1) and 1 on (2) and 2, and the top index (0) on (1). Thus for the initial condition in the second coordinate direction $x_2$ we have $v_H^{(2)}(0, x) = v^{(1)}(t_*, x)$.

Each function $v_{n+1}(t, x)$ in Eqs. (3.89) and (3.90) in a quasi-one-dimensional case or $v^{(1)}$ in Eqs. (3.91) and (3.92) is a solution of the linear equation that is a rather important feature of this algorithm. The algorithm comes from an approximation Newton-Kantorovich's method [8] in functional space.

Corresponding restrictions on sizes $\frac{\partial f}{\partial v}, \frac{\partial f}{\partial \dot{v}}$ and others are considered later. In order to reduce further records we will introduce notations:

$$f_j = c\dot{v}^{(j)} - b\frac{\partial v^{(j)}}{\partial x_j} - Y_j, \quad j = 1, 2, 3; \tag{3.93}$$

$$\begin{aligned}
&Y_1 = \sigma_1 a_3, \quad Y_2 = F_2 + \sigma_2 a_3, \quad Y_3 = F_3 + \sigma_3 a_3, \\
&F_2 = a_1 (T^{(1)})^k, \quad F_3 = a_2 \exp(A_4 T^{(2)}), \quad \frac{\partial T}{\partial v} = \frac{A_H}{A}, \\
&\frac{\partial f}{\partial \dot{v}} = c, \quad \frac{\partial f}{\partial v'} = -b, \quad \frac{\partial f_j}{\partial v} = -\phi_j, \\
&\phi_1 = 0, \quad R_1 = \sigma_1 a_3, \quad \phi_2 = kF_2 A_H/(AT^{(1)}), \quad A = s + zT^m, \\
&R_2 = F_2[1 - vkA_H/(AT^{(1)})] + \sigma_2 a_3, \quad \phi_3 = A_4 F_3 A_H/A, \\
&R_3 = F_3(1 - A_4 v A_H/A) + \sigma_3 a_3, \quad \sigma_j = 1/3, \quad j = 1, 2, 3.
\end{aligned} \tag{3.94}$$

Let's find the solution of the first boundary problem (3.91) and (3.92) at first in a coordinate direction $x_1$, using Eqs. (3.93), (3.83), and (3.84) with an index (1) above. We will substitute the last in a boundary problem (3.91) and (3.92) and for simplicity of further calculations we will introduce the right-hand side of the received equation through $h_1 = c\dot{v}^{(1)} - R_1(v_n)$. Then it will be:

$$\frac{\partial^2 v^{(1)}}{\partial x_1^2} = -b\frac{\partial v^{(1)}}{\partial x_1} - v^{(1)}\phi_1 + h_1. \tag{3.95}$$

Let's apply the Laplace integral transformation to the differential equation (3.95), excluding derivative on $x_1$ and replace it with its linear expression concerning the image of required function. Then we consider the functions for which the Laplace integral transformation converges absolutely. The valid part of complex number $p = \xi + i\eta, i = \sqrt{-1}$ is considered positive, that is Re $p > 0$. Let's introduce capital letters for images $V, H_1$ and omit an index (1) above. We have:

$$V(p, t, x_2, x_3) = \int_0^\infty \exp(-px_1)v(t, x)\, dx_1, v(t, x) = L^{-1}[V(p, t, x_2, x_3)],$$
$$h_1(t, x) = L^{-1}[H_1(p, t, x_2, x_3)].$$

We assume that the desired solution $v(t, x)$, and also its derivatives entering into the Eq. (3.95), satisfy Laplace integral transformation conditions on $x$, moreover growth degree on $x_1$ of the function $v(t, x)$ and its derivatives do not depend on $x_2$, $x_3$. Multiplying both parts of the Eq. (3.95) on $\exp(-px_1)$ and integrating on $x_1$ from 0 to $\infty$, we will receive [6] at $g_1 = v^{(1)}(t, 0, x_2, x_3), \left.\frac{\partial v^{(1)}}{\partial x_1}\right|_{x_1=0} = \frac{\partial g_1}{\partial x_1}$:

$$p^2 V(t, p, x_2, x_3) - pg_1(t, 0, x_2, x_3) - \frac{\partial g_1(t, 0, x_2, x_3)}{\partial x_1}$$
$$+ b[pV(t, p, x_2, x_3) - g_1(t, 0, x_2, x_3)] + \phi_1 V(t, p, x_2, x_3)$$
$$= H_1(t, p, x_2, x_3), \quad 0 < x_j < L_j, \quad j = 2, 3;$$

$$V = \frac{(p+\delta)g_1}{(p+\delta)^2 + b_1^2} + \frac{b_1\left(\delta g_1 + \frac{\partial g_1}{\partial x_1} + H_1\right)}{b_1[(p+\delta)^2 + b_1^2]}, \tag{3.96}$$

where $\delta = b/2, b_1 = \sqrt{\phi_1 - \delta^2}$.

Using the return Laplace integral transformation [6]: $L^{-1}[p/(p^2+b_1^2)] = \cos(b_1 x_1)$ at $b_1^2 > 0, L^{-1}[p/(p^2 - b_1^2)] = ch(b_1 x_1)$ at $b_1^2 < 0; L^{-1}[(p + \delta)^{-1}] = \exp(-\delta x_1), L^{-1}[H_1(p)/p] = \int_0^{x_1} h_1(y)\, dy$, we restore the original for $v(t, x)$ from Eq. (3.96) omitting, for brevity, the arguments $x_2, x_3$.

$$v(t, x) = \exp(-\delta x_1)\left\{g_1[u_1(x_1) + \delta u_2(x_1)] + u_2(x_1)\frac{\partial g_1}{\partial x_1}\right.$$
$$\left.+ \int_0^{x_1} \exp(\delta y) u_2(x_1 - y) h_1(y)\, dy\right\}, \quad 0 < x_j < L_j, \quad j = 2, 3, \tag{3.97}$$

$u_1(x_1) = \cos(b_1 x_1), u_2(x_1) = b_1^{-1}\sin(b_1 x_1), u_2(x_1 - y) = b_1^{-1}\sin[b_1(x_1 - y)]$ at $b_1^2 = \phi_1 - \delta^2 > 0$ from Eq. (3.96); $u_1(x_1) = ch(b_1 x_1), u_2(x_1) = b_1^{-1} sh(b_1 x_1), u_2(x_1 - y) = b_1^{-1} sh[b_1(x_1 - y)]$ at $b_1^2 < 0$.

Derivative $\frac{\partial g_1}{\partial x_1}$ in the expression (3.97) we will find using the second boundary condition in the first coordinate direction $x_1$ from Eq. (3.84), which is:

$$q_1(t, L_1, x_2, x_3) = \exp(-\delta L_1)\left\{g_1[u_1(L_1) + \delta u_2(L_1)] + u_2(L_1)\frac{\partial g_1}{\partial x_1} + \int_0^{L_1} \exp(\delta y)u_2(L_1 - y)h_1\, dy\right\},\ 0 < x_j < L_j,\ j = 2, 3. \tag{3.98}$$

Therefore finding $\frac{\partial g_1}{\partial x_1}$ in expression (3.98) and substituting it in the Eq. (3.97), we will have for $v$:

$$v(t,x) = \exp(-\delta x_1)\left\{g_1[u_1(x_1) + \delta u_2(x_1)] + \int_0^{x_1} \exp(\delta y)u_2(x_1 - y)h_1\, dy\right\} + \frac{u_2(x_1)}{u_2(L_1)}\exp(-\delta x_1)\left\{q_1 \exp(\delta L_1) - g_1[u_1(L_1) + \delta u_2(L_1)] - \int_0^{L_1} \exp(\delta y)u_2(L_1 - y)h_1\, dy\right\},\quad 0 < x_j < L_j,\quad j = 2, 3. \tag{3.99}$$

Let's transform the expression on the right-hand side (3.99) to get rid of the first integral with a variable top limit. Then entering Green's function $G(x_1, y)$ [44]

$$G(x_1, y) = \begin{cases} \frac{\exp[\delta(y - x_1)]}{u_2(L_1)}[u_2(x_1 - y)u_2(L_1) - u_2(x_1) \times u_2(L_1 - y)], \\ \quad 0 \le y \le x_1; \\ -\exp[\delta(y - x_1)]u_2(x_1)u_2(L_1 - y)/u_2(L_1), \quad x_1 \le y \le L_1, \end{cases} \tag{3.100}$$

expression (3.99) using formulas (3.94), it will be rewritten:

$$\dot{v}^{(1)} - U_1 v^{(1)} = Z_1^{-1}\left[S(x_1) + \int_0^{L_1} G(x_1, y)R_1(v^{(0)})\, dy\right] = W_1(v^{(0)}, x, t),$$
$$U_1 = Z_1^{-1},\quad S(x_1) = \exp(-\delta x_1)u_2(x_1)\{[u_1(L_1) + \delta u_2(L_1)]g_1 - q_1 \exp(\delta L_1)\}/u_2(L_1) - \exp(-\delta x_1)[u_1(x_1) + \delta u_2(x_1)]g_1,$$
$$Z_1 = \int_0^{L_1} c(v^{(0)})G(x_1, y)\, dy,\quad v^{(1)}(0, x) = v_H(x),$$
$$0 < x_j < L_j, j = 2, 3. \tag{3.101}$$

As a result, the problem solution (3.101) will be [26] [$t = t_*$ from Eqs. (3.73) and (3.76)]:

$$v^{(1)}(x,t_*) = \left\{ v_H^{(1)} + \int_0^{t_*} W_1(v^{(0)},x,\tau)\exp[-\tau U_1(v^{(0)})]\,d\tau \right\}$$
$$\times \exp[t_* U_1(v^{(0)})], \quad v_H^{(1)} = v_H(x), \quad v^{(0)} = v_n,$$
$$n = 0,1,2,\ldots, \tag{3.102}$$
$$T^{(1)} = T^{(0)} - f(T^{(0)})/f_T'(T^{(0)}), \quad f_T'(T^{(0)}) = \frac{\partial f(T^{(0)})}{\partial T^{(0)}},$$
$$f(T^{(0)}) = z(T^{(0)})^{m+1} + s(m+1)T^{(0)} - A_H(m+1)v^{(1)}(t_*,x),$$
$$f_T'(T^{(0)}) = z(m+1)(T^{(0)})^m + s(m+1),$$

The same solutions can be for boundary problems (3.81) and (3.82) using Eqs. (3.89) and (3.90) or Eqs. (3.91) and (3.92) and other equations from Eqs. (3.85)–(3.88) and formulas (3.94) in coordinate directions $x_2, x_3$. Then according to formulas (3.72)–(3.78) we have:

$$v^{(j)}(x,t_*) = \left[ v_H^{(j)} + \int_0^{t_*} W_j(v^{(j-1)},x,\tau)\exp(-U_j\tau)\,d\tau \right]$$
$$\times \exp(t_* U_j), \quad v_H^{(j)} = v^{(j-1)}, \quad j = 2,3, \tag{3.103}$$
$$W_j = Z_j^{-1}\left[ S(x_j) + \int_0^{L_j} G(x_j,y)R_j(v^{(j-1)})\,dy \right],$$
$$U_j = Z_j^{-1}, \quad S(x_j) = \exp(-\delta x_j)u_2(x_j)\{[u_1(L_j) + \delta u_2(L_j)]g_j$$
$$- q_j\exp(\delta L_j)\}/u_2(L_j) - \exp(-\delta x_j)[u_1(x_j) + \delta u_2(x_j)]g_j,$$
$$Z_j = \int_0^{L_j} c(v^{(j-1)})G(x_j,y)\,dy, \quad n = 0,1,2,\ldots,$$
$$T^{(j)} = T^{(j-1)} - f(T^{(j-1)})/f_T'(T^{(j-1)}), \quad f_T'(T^{(j-1)}) = \frac{\partial f(T^{(j-1)})}{\partial T^{(j-1)}},$$
$$f(T^{(j-1)}) = z(T^{(j-1)})^{m+1} + s(m+1)T^{(j-1)} - A_H(m+1)v^{(j)}(t_*,x),$$
$$f_T'(T^{(j-1)}) = z(m+1)(T^{(j-1)})^m + s(m+1), \quad j = 2,3.$$

Let's note that $G(x_j, y), j = 2, 3$ comes from Eq. (3.100) by replacement of arguments $x_1, L_1$ in accordance with a sequence on $x_j, L_j, j = 2, 3$. At $x = x_2$ in Eq. (3.103) other variables $0 < x_j < L_j, j = 1, 3$ are changed parametrically, as in Eq. (3.102). The same situation is for $x = x_3$, thus the final solution of the first boundary problem will be (3.81) and (3.82):

$v^{(3)}(t_*, x) = v_{n+1}(t_*, x), \forall t_* > 0, n = 0, 1, 2, \ldots$, and in the presence of the inversion formula (3.67) of the initial nonlinear boundary problem (3.60)–(3.62).

## Existence, Uniqueness, and Convergence

We will consider the nonlinear case $m > 0$, $k \geq 2$ and for simplicity of the analysis range of definition $Q : (0 \leq x_j \leq b,\ b = \min(L_j), j = 1, 2, 3)$, $0 \leq t \leq t_k$ at $A_2 = A_3 = B = 0, C = 1, A_1 = -1$.

Then using the Kirchhoff's transformation (3.64) and formulas (3.65) for (3.60)–(3.62) at zero boundary conditions of the 1st type we have:

$$\sum_{j=1}^{3} \frac{\partial^2 v}{\partial x_j^2} = \frac{1}{A}\frac{\partial v}{\partial t} + (T(v))^k, \quad v(0, x) = v_H(x), \quad v|_\Gamma = 0, \tag{3.104}$$

a boundary problem (3.104), at $A = s + zw^m$, and $v_H$ is possible to find from the inversion formula (3.68). As a result of algorithm application (3.89)–(3.99), (3.101)–(3.103) the solution of a boundary problem (3.104) is written in coordinate directions $x_j, j = 1, 2, 3$ (in the one-dimensional case the solution is received in [15]):

$$\begin{aligned}
v^{(j)} &= \left\{ v_H^{(j)} + \int_0^{t_*} W_j \exp[\tau U_j(v^{(j-1)})]\, d\tau \right\} \exp[-t_* U_j(v^{(j-1)})],\\
&j = 1, 2, 3, v^{(0)} = v_n, \quad n = 0, 1, 2, \ldots, \qquad (3.105)\\
&v_H^{(1)} = v_H(x), \quad v_H^{(j)} = v^{(j-1)}, \quad j = 2, 3, \quad v^{(3)} = v_{n+1}(t_*, x);\\
U_j(v^{(j-1)}) &= Z_j^{-1}\left[\int_0^b G(x_j, y)\phi_j(v^{(j-1)})\, dy - 1\right],\\
Z_j &= \int_0^b (G(x_j, y)/A)\, dy, \quad W_j = \int_0^b R_j(v^{(j-1)}) G(x_j, y)\, dy,\\
T^{(j)} &= T^{(j-1)} - f(T^{(j-1)})/f_T'(T^{(j-1)}), \quad f_T'(T^{(j-1)}) = \frac{\partial f(T^{(j-1)})}{\partial T^{(j-1)}},\\
f(T^{(j-1)}) &= z(T^{(j-1)})^{m+1} + s(m+1)T^{(j-1)} - A_H(m+1)v^{(j)}(t_*, x),\\
f_T'(T^{(j-1)}) &= z(m+1)(T^{(j-1)})^m + s(m+1), \quad j = 1, 2, 3;\\
W_j &= \phi_j = 0, \quad j = 2, 3,
\end{aligned}$$

and $G(x_j, y), \phi_1 = \phi_1(F_2), R_1 = R_1(F_2)$ are defined from Eqs. (3.100) and (3.94).

Using formulas (3.76)–(3.80) the final solution from Eq. (3.105) will be (an index $*$ below at $t$ will be omitted):

$$v^{(3)}(t,x) = \exp[-tU(v_n)]\left\{v_H + \int_0^t W_1(v_n)\exp[\tau U_1(v_n)]\,d\tau\right\},$$
$$U(v_n) = \sum_{j=1}^{3} U_j(v^{(j-1)}), \quad v^{(3)}(t,x) = v_{n+1}(t,x), \quad n = 0,1,2,\ldots. \tag{3.106}$$

Let's say that in some neighborhood of a root function $f = f(v,\dot{v})$ from Eq. (3.89) together with the partial derivatives $\frac{\partial f}{\partial v}, \frac{\partial^2 f}{\partial v^2}, \frac{\partial f}{\partial \dot{v}}, \frac{\partial^2 f}{\partial \dot{v}^2}$ is continuous, and in this neighborhood $\frac{\partial f}{\partial v}, \frac{\partial^2 f}{\partial v^2}, \frac{\partial f}{\partial \dot{v}}, \frac{\partial^2 f}{\partial \dot{v}^2}$ does not go to zero.

Existence and uniqueness of the solution (3.106) of a boundary problem (3.104) are proved in the same manner as was made in Chapter 2 [see formulas (2.19)–(2.30)].

An estimation of speed of convergence. Let's use locally-one-dimensional scheme splitting (3.83)–(3.88) to a boundary problem (3.104) in such a way as it was made for system (3.81) and (3.82) so we have $\sigma = 1/3$

$$\frac{\partial^2 v^{(j)}}{\partial x_j^2} = \frac{\dot{v}^{(j)}}{A^{(j-1)}} + F_j, \quad j = 1,2,3, \tag{3.107}$$
$$v^{(1)}(0,x) = v_H(x), \quad v^{(j)}(0,x) = v^{(j-1)}(t_*,x), \quad j = 2,3,$$
$$v^{(j)}|_\Gamma = 0, \quad F_j = \sigma a_1[T^{(j-1)}(v)]^k, \quad j = 1,2,3.$$

Let's address a recurrence relationship (3.89) and, noticing that $f(v,\dot{v}) = s(v) + r(\dot{v})$, we will subtract $n$-e the equation from $(n+1)$th which corresponds to the first equation (3.91) for $v^{(1)}$ in a quasi-one-dimensional variant, then we receive on the coordinate direction $x_1$:

$$\frac{\partial^2(v^{(1)} - v_n)}{\partial x_1^2} = s(v_n) - s(v_{n-1}) - (v_n - v_{n-1})\frac{\partial s(v_{n-1})}{\partial v}$$
$$+ (v^{(1)} - v_n)\frac{\partial s(v_n)}{\partial v} + r(\dot{v}_n) - r(\dot{v}_{n-1})$$
$$- (\dot{v}_n - \dot{v}_{n-1})\frac{\partial r(\dot{v}_{n-1})}{\partial \dot{v}} + (\dot{v}^{(1)} - \dot{v}_n)\frac{\partial r(\dot{v}_n)}{\partial \dot{v}}. \tag{3.108}$$

From the average theorem [33] it follows:

$$s(v_n) - s(v_{n-1}) - (v_n - v_{n-1})\frac{\partial s(v_{n-1})}{\partial v} = 0.5(v_n - v_{n-1})^2\frac{\partial^2 s(\xi)}{\partial v^2},$$
$$v_{n-1} \le \xi \le v_n.$$

Let's consider Eq. (3.108) how the equation is relative to $u^{(1)} = v^{(1)} - v_n, (u^{(0)} = u_n, u_n = v_n - v_{n-1})$, transform it as above Eqs. (3.96)–(3.99), (3.101), and (3.103), so we have:

$$\dot{u}^{(1)} - u^{(1)}\left[1 - \int_0^b G(x_1, y)\left(\frac{\partial s(v_n)}{\partial v}\right) dy\right] Z_1^{-1} = -0.5Z_1^{-1}$$

$$\times \int_0^b G(x_1, y)\left[u_n^2 \frac{\partial^2 s(v_n)}{\partial v^2} + \dot{u}_n^2 \frac{\partial^2 r(\dot{v}_n)}{\partial \dot{v}^2}\right] dy,$$

$$u_H = 0, \quad Z_1 = \int_0^b G(x_1, y)\left[\frac{\partial r(\dot{v}_n)}{\partial \dot{v}}\right] dy, \quad \dot{u}_n = \dot{u}^{(0)}. \tag{3.109}$$

As a result, the problem solution (3.109) on the first coordinate direction $x_1$ will look like Eq. (3.102), where $v_H = 0$. In the same manner it is possible to receive solutions of a type (3.103) on coordinate directions $x_2, x_3$ from Eq. (3.107). Final solutions like Eqs. (3.105) and (3.106) using formulas (3.76)–(3.80) will be rewritten as:

$$u^{(3)}(x, t_*) = \exp(t_* U) \int_0^{t_*} W_1 \exp(-\tau U_1)\, d\tau + \exp[t_*(U_2 + U_3)]$$

$$\times \int_0^{t_*} W_2 \exp(-\tau U_2) d\tau + \exp(t_* U_3) \int_0^{t_*} W_3 \exp(-\tau U_3)\, d\tau, \tag{3.110}$$

$$U = \sum_{j=1}^{3} U_j(v^{(j-1)}), \quad U_j(v^{(j-1)}) = Z_j^{-1}\left\{1 - \int_0^b G(x_j, y)\right.$$

$$\left.\times \left[\frac{\partial s(v^{(j-1)})}{\partial v}\right] dy\right\}, \quad W_j = -0.5Z_j^{-1} \int_0^b G(x_j, y)\left[(u^{(j-1)})^2\right.$$

$$\left.\times \frac{\partial^2 s(v^{(j-1)})}{\partial v^2} + (\dot{u}^{(j-1)})^2 \frac{\partial^2 r(\dot{v}^{(j-1)})}{\partial \dot{v}^2}\right] dy, \quad u^{(0)} = u_n,$$

$$Z_j = \int_0^b G(x_j, y)\left[\frac{\partial r(\dot{v}^{(j-1)})}{\partial \dot{v}}\right] dy, \quad j = 1, 2, 3, \quad n = 0, 1, 2, \ldots.$$

Let's put $\max\limits_{v,\dot{v}\in R}\left(\left|\frac{\partial s(v^{(j)})}{\partial v}\right|, \left|\frac{\partial r(\dot{v}^{(j)})}{\partial \dot{v}}\right|\right) = c_1, \max\limits_{x_j, y}|G(x_j, y)| = b/4$ [7], $\max\limits_{v,\dot{v}\in R}\left(\left|\frac{\partial^2 s(v^{(j)})}{\partial v^2}\right|, \left|\frac{\partial^2 r(\dot{v}^{(j)})}{\partial \dot{v}^2}\right|\right) = c_2, j = 0, 1, 2$, assuming $c_p < \infty, p = 1, 2$. Then, noting that $\frac{\partial^2 r(\dot{v}^{(j)})}{\partial \dot{v}^2} = 0, j = 0, 1, 2$ and using an assumption about equality of all directions in space ($U_j = \alpha, W_j = Bu_n^2, j = 1, 2, 3$) and an equality of functions $u^{(0)} = u^{(j)}, j = 1, 2$ (for converging sequence $v_n$ all intermediate values $u^{(j)}, j = 0, 1, 2$ are close to zero as they are in a convergence interval: $[v^{(0)}, v^{(3)}]$), we have from Eqs. (3.109) and (3.83) at $u_{n+1}(t_*, x) = u^{(3)}(t_*, x)$, omitting an index $(*)$ at $t$ below:

$$|u_{n+1}| \le Bu_n^2 V \exp(\alpha t) \int_0^t \exp(-\alpha \tau)\, d\tau,$$

$$V = \exp(2\alpha t) + \exp(\alpha t) + 1, \quad B = c_2/2c_1, \quad \alpha = 4/(c_1 b^2) - 1,$$

$$|u_{n+1}| \le Bu_n^2[\exp(\alpha t) - 1]V/\alpha. \tag{3.111}$$

Let's choose $u_0(t, x)$ so that $|u_0(t, x)| \le 1$ in region Q. As a result from the expression (3.84) at $n = 0$ we receive introducing $M_1 = \max_Q |u_1|$:

$$M_1 \le B[\exp(3\alpha t) - 1]/\alpha = S. \tag{3.112}$$

Hence, under condition $\alpha > 0 (b < 2/\sqrt{c_1})$ we find that the top border $M_1$ will not surpass 1, if there is inequality $S \le 1$ in Eq. (3.112):

$$t \le \ln(\alpha/B + 1)^{1/3\alpha}. \tag{3.113}$$

Therefore choosing intervals $[0, t], [0, b]$ small enough so that a condition is satisfied from Eq. (3.113), we will have $M_1 \le 1$. Finally we find by induction:

$$\max_{x,t \in Q} |v_{n+1} - v_n| \le S \max_{x,t \in Q} |v_n - v_{n-1}|^2. \tag{3.114}$$

The relationship (3.114) shows that if a convergence in general takes place, it is quadratic. Thus, enough big $n$ each following step doubles a number of correct signs in the given approximation.

## Results of Test Checks

Accuracy of the received approximate analytical formulas (3.102) and (3.103) we will check practically while solving the boundary problem for the equation in partial derivatives. For simplicity of the analysis we will consider boundary conditions of the 1st type in the region $Q : (0 \le x_j \le b, j = 1, 2, 3), 0 \le t \le t_k$:

$$A_6 \frac{\partial T}{\partial t} = \operatorname{div}(A_2 \nabla T) + A_1 \nabla T + A_3 T^k + A_4 \exp(A_5 T) + F(x, t), \tag{3.115}$$

$$T|_{t=0} = \exp(y_1 + y_2 + y_3), \quad y_j = x_j/b, \tag{3.116}$$

$$T|_{x_1=0} = \exp(t + y_2 + y_3), \quad T|_{x_2=0} = \exp(t + y_1 + y_3),$$
$$T|_{x_3=0} = \exp(t + y_1 + y_2), \quad T|_{x_j=b}$$
$$= \exp(1) \cdot T|_{x_j=0}, \quad j = 1, 2, 3. \tag{3.117}$$

The exact solution of a problem (3.115)–(3.117) was taken: $T = \exp(t + z), z = \sum_{j=1}^{3} y_j$, then a source $F$ in the Eq. (3.115) will be

$$F = \exp(t + z)\{A_6 - 3A_1/b - 3[s + (m + 1)\exp(m(t + z))]/b^2\}$$
$$- A_3 \exp[k(t + z)] - A_4 \exp[A_5 \exp(t + z)],\ A_2 = s + T^m,\ m \neq -1.$$

The following basic values of the initial data were used: $m = 0.5, k = 1, s = A_H = A_6 = 1, t = 1, b = 1; N_j = 11, \Delta x_j = b/(N_j - 1), j = 1, 2, 3, M = 101, \Delta t = t/(M - 1); N_j, M, \Delta x_j, j = 1, 2, 3, \Delta t$ is a number of checkouts steps on space and time while finding integrals in the Eqs. (3.101)–(3.103) by Simpson's formula [21].

The program is made in the language Fortran-90, calculation was made on Pentium 3 (800 MHz, compiler PS 5) with double accuracy. In Table 3.2 you can see the maximum relative error $\epsilon = |T - \tilde{T}|100\%/T$ ($T$ is exact, $\tilde{T}$ is approximate analytical solution) at various values $A_j, j = 1, 3, 4, 5$ for a basic variant.

The boundary problem (3.115)–(3.117) was solved by means of formulas (3.102) and (3.103). Number of iterations ($J$) was traced [for total expressions of the type (3.106) and (3.109) according to formulas (3.76)–(3.80)] on relative change of an error vector (in percentage):

$$||V_n|| = \max_{x,t \in Q} |(v_{n+1} - v_n)/v_{n+1}|.$$

In Table 3.2 there are results of test calculations $||V_n|| \leq \delta,\ \delta = 1\%$. Thus calculation time of any variant is $t_p = 5$ s.

In comparison with the numerical solution of a problem (3.115)–(3.117) locally-one-dimensional scheme type splittings (3.72)–(3.75) were

**Table 3.2** A dependence of the maximum relative errors at the various values $A_j, j = 1, 3, 4, 5$

| N | 1 | 2 | 3 | 4 | 5 | 6 | 7 |
|---|---|---|---|---|---|---|---|
| $A_1$ | 0 | 1 | −1 | 1 | 1 | 1 | 1 |
| $A_3$ | 0 | 0 | 0 | 1 | −1 | 0 | 1 |
| $A_4$ | 0 | 0 | 0 | 0 | 0 | 1 | 1 |
| $A_5$ | 0 | 0 | 0 | 0 | 0 | −1 | −1 |
| $\varepsilon, \%$ | 1.24 | 1.51 | 1.38 | 1.76 | 2.0 | 1.51 | 1.76 |
| $J$ | 2 | 2 | 2 | 2 | 2 | 2 | 2 |

used. For the solution of the quasi-one-dimensional equations from system (3.83)–(3.88) the scheme of implicit absolutely steady difference with approximation error used (in total sense [12]) for the first and second derivative on space is $O\left[\sum_{j=1}^{3}\Delta x_j^2\right]$ and the two-layer scheme for derivative on time with a margin error approximations is $O(\Delta t)$. Thus on each step on time a Pikar's method of consecutive approximation that converged for two iterations was used.

For basic initial sizes and $\tau = 0.01$ the numerical solution of a variant at number 3 from Table 3.2 gives $\epsilon = 2.75\%$ to time moment $t = 1$ and $t_p = 5$ s. However, in conditions of practical work, a number of nodes of difference grids on space as a rule is $N \sim 21 \div 41$. It leads to that in three-dimensional case time $t_p$ can increase: $t_p = 25$ s $(N = 21, \tau = 0.005)$, $t_p =$ 5.5 min. $(N = 41, \tau = 0.002)$.

It is of interest to receive the solution of a boundary problem (3.115)–(3.117) for real values of initial data, for example at modeling of convective-conductive heat-transfer in a problem of thermal protection [31], where the geometrical sizes of range of definition of a problem (a thickness) are 1 m. In this case there is less than heat-shielding covering at $b = 10^{-2}$ m; we take a material of copper [31]: $A_6 = 9 \cdot 10^3$ s/m$^2$, and in the calculation results in Table 3.3 we can see the basic initial data.

Thus, in the presence of the inversion formula (3.67) in the wide range of change of the initial data $10^{-2} \le b \le 1, 1 \le A_6 \le 9 \cdot 10^3, |A_1| \le 400$ we receive the estimation of approximate analytical solution of modeling nonlinear boundary problems (3.115)–(3.117) by means of method quasi-linearization, locally-one-dimensional scheme splittings and Laplace integral transformation. Also, on a concrete example the estimation of speed of convergence of iterative process is received and it is proved that this size decreases with growth $n$ and by the quadratic law.

**Table 3.3** A dependence of the maximum relative errors at the various values $A_j, j = 1, 3, 4, 5$

| N | 1 | 2 | 3 | 4 | 5 | 6 | 7 | 8 |
|---|---|---|---|---|---|---|---|---|
| $k$ | 1 | 1 | 1 | 1 | 1 | 1 | 2 | 1 |
| $A_1$ | 0 | 1 | −1 | −400 | 400 | 1 | 1 | 1 |
| $A_3$ | 0 | 0 | 0 | 0 | 0 | 1 | 1 | 0 |
| $A_4$ | 0 | 0 | 0 | 0 | 0 | 0 | 0 | 1 |
| $A_5$ | 0 | 0 | 0 | 0 | 0 | 0 | 0 | −1 |
| $\varepsilon$, % | 1.6 | 1.6 | 1.6 | 1.3 | 5.65 | 1.6 | 1.6 | 1.6 |
| $J$ | 2 | 2 | 2 | 2 | 2 | 2 | 2 | 2 |

## 3.3 METHOD OF SOLUTION OF THREE-DIMENSIONAL NONLINEAR BOUNDARY PROBLEMS FOR PARABOLIC EQUATION OF GENERAL TYPE

### Statement of a Problem and a Method Algorithm

Let us attempt to find a solution of nonlinear equation of parabolic type with the elliptic operator containing mixed derivatives [41], with sources [2, 5]:

$$C(T)\frac{\partial T}{\partial t} = \sum_{j=1}^{3}\frac{\partial}{\partial x_j}\left[A(T)\frac{\partial T}{\partial x_j}\right] + B(T)\sum_{j=1}^{3}\frac{\partial T}{\partial x_j}$$
$$+ w\sum_{j=1, i\neq j}^{3}\frac{\partial}{\partial x_j}\left[A(T)\frac{\partial T}{\partial x_i}\right]$$
$$+ A_1T^k + A_2\exp(A_4T) + A_3(x,t) \tag{3.118}$$

in a parallelepiped $Q : x = (x_1, x_2, x_3), (0 < x_j < L_j; 0 < L_j < \infty, j = 1, 2, 3)$ at $0 < t \leq t_k, A_j = \text{const}, j = 1, 2, 4, w = \text{const}$ with the initial condition

$$T|_{t=0} = T_H(x) \tag{3.119}$$

with boundary value of a type

$$-A\frac{\partial T}{\partial \bar{n}}|_\Gamma = \gamma(T|_\Gamma - \theta), \tag{3.120}$$

where $\gamma = \text{const}$, $\bar{n}$ is a vector of external normals to the limiting surfaces $\Gamma$ initial range of definition Q, $\theta$ is a value in environment.

At $\gamma = 0$ we receive Neumann's condition, and at $\gamma \to \infty$ we receive Dirichlet condition (boundary condition of the 1st type)

$$T|_\Gamma = \Psi, \tag{3.121}$$

where $\Psi \neq \text{const}$.

The condition of parabolicity for the Eq. (3.118) looks like [12]: $|w| < 1$.

The equation of a type (3.118) is applied in mechanics of reacting environments [2].

Then using the inversion formula, we take

$$A(T) = s + zT^m,$$

where $m > -1$, $s > 0$, $z$ are constants [5].

Let's use the Kirchhoff's transformation [5]

$$v = \int_0^T \frac{A(T)}{A_H}\, dT, \tag{3.122}$$

where $A_H$ is for example, the heat conductivity factor at the temperature equal to zero. Then taking into consideration the relationships [5]:

$$\nabla A = \frac{\partial A}{\partial T}\nabla T, \quad \frac{\partial v}{\partial t} = \frac{A}{A_H}\frac{\partial T}{\partial t}, \quad \nabla v = \frac{A}{A_H}\nabla T, \tag{3.123}$$

we find the differential equation from Eqs. (3.118) and (3.123)

$$\nabla^2 v = c(v)\frac{\partial v}{\partial t} - b(v)\nabla v - w\sum_{j=1, i\neq j}^{3}\frac{\partial^2 v}{\partial x_j \partial x_i}$$
$$- a_1 T^k - a_2 \exp(A_4 T) - a_3(x, t), \tag{3.124}$$

where $c = C/A, b = B/A, a_j = A_j/A_H, j = 1, 2, 3, A_4, a_1, a_2$ are constants.

Then we have from the formula $A(T) = s + zT^m$ and equalities (3.122), applying the Newton's method [7]:

$$T_{n+1} = T_n - \frac{f(T_n)}{f_T'(T_n)}, \quad f_T'(T_n) = \frac{\partial f(T_n)}{\partial T_n},$$
$$f(T_n) = zT_n^{m+1} + s(m+1)T_n - A_H(m+1)v(t, x),$$
$$f_T'(T_n) = z(m+1)T_n^m + s(m+1), \quad n = 0, 1, 2, \ldots. \tag{3.125}$$

Here as initial approximation $T_0 = \text{const}$ we take any constant number close to $T_H$, and $v$ is defined below.

Let's note that variation ranges of independent variables and type of the boundary conditions do not change in relation to Kirhhoff's transformation (3.122), and within the inversion formula (3.125) the boundary conditions of the 1st, 2nd, and 3rd type passes into Dirichlet, Neumann, and Newton's conditions. Using formulas (3.122) and (3.123) initial and boundary conditions (3.119), (3.120), and (3.121) for the Eq. (3.124) will be rewritten:

$$v|_{t=0} = v_H, \quad v_H = [sT_H + zT_H^{m+1}/(m+1)]/A_H = F(T_H), \tag{3.126}$$

$$-A_H\frac{\partial v}{\partial \bar{n}}|_\Gamma = \gamma(T(v)|_\Gamma - \theta), \tag{3.127}$$

where $T(v)$ is generally defined from Eq. (3.125).

At $\gamma \to \infty$ we receive a boundary condition of the 1st type from Eq. (3.127):

$$v|_\Gamma = A_H^{-1}[sT + zT^{m+1}/(m+1)]|_\Gamma = \Phi. \tag{3.128}$$

Let's apply locally-one-dimensional scheme splittings to Eq. (3.124) at the differential level [12] and for simplicity of the analysis, we will consider the mixed boundary problem

$$\sum_{j=1}^{3} \frac{\partial^2 v}{\partial x_j^2} = c\frac{\partial v}{\partial t} - b\sum_{j=1}^{3} \frac{\partial v}{\partial x_j} - w \sum_{j=1, i\neq j}^{3} \frac{\partial^2 v}{\partial x_j \partial x_i} - a_1 T^k$$
$$- a_2 \exp(A_4 T) - a_3(x,t),$$
$$v|_{t=0} = F(T_H), \quad -A_H \frac{\partial v}{\partial \bar{n}}|_\Gamma = \gamma(T|_\Gamma - \theta). \qquad (3.129)$$

At $\gamma \to \infty$ we receive a boundary condition of the 1st type from Eq. (3.128)

$$v|_\Gamma = \Phi. \qquad (3.130)$$

As a result we have for boundary conditions of the mixed type (1st–3rd types) from Eqs. (3.129) and (3.130):

$$\frac{\partial^2 v^{(1)}}{\partial x_1^2} = c\frac{\partial v^{(1)}}{\partial t} - b\frac{\partial v^{(1)}}{\partial x_1} - a_1 (T^{(1)})^k - \sigma_1 a_3, \quad 0 < t < t_*; \qquad (3.131)$$

$$v^{(1)}(0,x) = v_H(x),$$
$$\left.\frac{\partial v^{(1)}}{\partial x_1}\right|_{x_1=0} = -\gamma_0(T^{(1)}|_{x_1=0} - \theta_0)/A_H,$$
$$\left.\frac{\partial v^{(1)}}{\partial x_1}\right|_{x_1=L_1} = -\gamma_{L_1}(T^{(1)}|_{x_1=L_1} - \theta_{L_1})/A_H; \qquad (3.132)$$

$$\frac{\partial^2 v^{(2)}}{\partial x_2^2} = c\frac{\partial v^{(2)}}{\partial t} - b\frac{\partial v^{(2)}}{\partial x_2} - 2w\frac{\partial^2 v^{(2)}}{\partial x_1 \partial x_2}$$
$$- a_2 \exp(A_4 T^{(2)}) - \sigma_2 a_3, \quad 0 < t < t_*; \qquad (3.133)$$

$$v^{(2)}(0,x) = v^{(1)}(t_*,x), \quad v^{(2)}|_{x_2=0} = g_2(t,x_1,x_3),$$
$$v^{(2)}|_{x_2} = L_2 = q_2(t,x_1,L_2,x_3); \qquad (3.134)$$

$$\frac{\partial^2 v^{(3)}}{\partial x_3^2} = c\frac{\partial v^{(3)}}{\partial t} - b\frac{\partial v^{(3)}}{\partial x_3} - 2w\left(\frac{\partial^2 v^{(3)}}{\partial x_1 \partial x_3} + \frac{\partial^2 v^{(3)}}{\partial x_2 \partial x_3}\right)$$
$$- \sigma_3 a_3, \quad 0 < t < t_*; \qquad (3.135)$$

$$v^{(3)}(0,x) = v^{(2)}(t_*,x), \quad v^{(3)}|_{x_3=0} = g_3(t,x_1,x_2),$$
$$v^{(3)}|_{x_3=L_3} = q_3(t,x_1,x_2,L_3), \qquad (3.136)$$

where $\sigma_1+\sigma_2+\sigma_3 = 1$, and $T(v)$ is defined from Eq. (3.125). Our purpose is to receive the solution of a nonlinear boundary problem if it exists as a limit of sequence of solutions of the linear boundary problems. For this purpose we will use the results of the work [7]. Let's assume further that all coordinate directions in space are equivalent.

Let $v_0 = \text{const}$ be some initial approximation [as an initial approximation it is better to take the value close to $v_H$ from Eq. (3.126)]. For simplicity of the analysis we will consider the quasi-one-dimensional case and the sequence $v_n(t,x)$, defining recurrence relationship [7] (a point above and a stroke on the right-hand side corresponds to a partial derivative on time and on space accordingly):

$$\frac{\partial^2 v_{n+1}}{\partial y^2} = f + (v_{n+1} - v_n)\frac{\partial f}{\partial v} + (v'_{n+1} - v'_n)\frac{\partial f}{\partial v'} + (v''_{n+1} - v''_n)\frac{\partial f}{\partial v''} + (\dot{v}_{n+1} - \dot{v}_n)\frac{\partial f}{\partial \dot{v}}, \quad f = f(v_n, \dot{v}_n, v'_n, v''_n), \tag{3.137}$$

$$v_H = v_n(0,x), \quad -A_H\frac{\partial v_{n+1}}{\partial y}|_\Gamma = \gamma(T|_\Gamma - \theta); \quad n = 0,1,2,\ldots, \tag{3.138}$$

where $\gamma$ is any of coordinates $x_j, j = 1,2,3$ in Eqs. (3.137) and (3.138). Then at $\gamma = x_1$ other coordinates in Eq. (3.137), $0 < x_j < L_j, j = 2,3$ are changed parametrically. On the remained coordinates while receiving expressions (3.137) there is a circular replacement of indexes when instead of $\gamma$ we substitute $x_2, x_3$ accordingly. We will notice that in solution of a three-dimensional boundary problem (3.129), if in the first coordinate direction $x_1$ as the initial iterations acts as $v_n$, according to formulas from Section 3.2 of this chapter (Eqs. 3.49–3.53), the subsequent iteration $v_{n+1}$ will be from definitive expression of type (3.53) in which it is necessary to put $v_{n+1}(t_*,x) = v^{(3)}(t_*,x)$ [see comment to the formula later (3.161)]. Then in the quasi-one-dimensional equation variant in Eqs. (3.137) and (3.138) on coordinate $x_1$ in accordance with Eqs. (3.131) and (3.132) it will be rewritten:

$$\frac{\partial^2 v^{(1)}}{\partial x_1^2} = f_1 + (v^{(1)} - v^{(0)})\frac{\partial f_1}{\partial v} + (v'^{(1)} - v'^{(0)})\frac{\partial f_1}{\partial v'} + (\dot{v}^{(1)} - \dot{v}^{(0)})\frac{\partial f_1}{\partial \dot{v}}, \quad v^{(0)} = v_n, \quad f_1 = f_1(v^{(0)}, \dot{v}^{(0)}, v'^{(0)}); \tag{3.139}$$

$$v_H^{(1)} = v_H(x), \quad \left.\frac{\partial v^{(1)}}{\partial x_1}\right|_{x_1=0} = -\gamma_0(T^{(1)}|_{x_1=0} - \theta_0)/A_H,$$

$$\left.\frac{\partial v^{(1)}}{\partial x_1}\right|_{x_1=L_1} = -\gamma_{L_1}(T^{(1)}|_{x_1=L_1} - \theta_{L_1})/A_H, \quad n = 0, 1, 2, \ldots . \quad (3.140)$$

It is Eqs. (3.139) and (3.140) which make it possible to write down the expressions like these on other coordinate directions $x_2, x_3$. In particular, for the second coordinate direction $x_2$ it is necessary in Eqs. (3.139) and (3.140) to replace the top and bottom indexes (1) and 1 on (2) and 2, and the top index (0) on (1). Thus for the initial condition in the second coordinate direction $x_2$ we have $v_H^{(2)}(0, x) = v^{(1)}(t_*, x)$.

Each function $v_{n+1}(t, x)$ in Eqs. (3.137) and (3.138) in a quasi-one-dimensional case or $v^{(1)}$ in Eqs. (3.139) and (3.140) is a solution of the linear equation that is a rather important feature of this algorithm. The algorithm comes from an approximation of Newton-Kantorovich's method [8] in functional space.

It is supposed that in some neighborhood of a root function $f = f(v, \dot{v}, v', v'')$ from Eq. (3.137) together with the partial derivatives $\frac{\partial f}{\partial v}, \frac{\partial f}{\partial v'}, \frac{\partial f}{\partial v''}, \frac{\partial f}{\partial \dot{v}}$ is continuous, and $\frac{\partial f}{\partial v}, \frac{\partial f}{\partial v'}, \frac{\partial f}{\partial v''}, \frac{\partial f}{\partial \dot{v}}$ in this neighborhood does not go to zero.

To reduce further records we will introduce notations:

$$f_1 = c\dot{v}^{(1)} - b\frac{\partial v^{(1)}}{\partial x_1} - Y_1, \quad f_2 = c\dot{v}^{(2)} - b\frac{\partial v^{(2)}}{\partial x_2} - 2w\frac{\partial^2 v^{(2)}}{\partial x_1 \partial x_2} - Y_2,$$

$$f_3 = c\dot{v}^{(3)} - b\frac{\partial v^{(3)}}{\partial x_3} - 2w\left(\frac{\partial^2 v^{(3)}}{\partial x_1 \partial x_3} + \frac{\partial^2 v^{(3)}}{\partial x_2 \partial x_3}\right) - Y_3; \quad (3.141)$$

$$Y_1 = F_1 + \sigma_1 a_3, \quad Y_2 = F_2 + \sigma_2 a_3, \quad Y_3 = \sigma_3 a_3,$$

$$F_1 = a_1 (T^{(1)})^k, \quad F_2 = a_2 \exp(A_4 T^{(2)}), \quad \frac{\partial T}{\partial v} = \frac{A_H}{A},$$

$$\frac{\partial f}{\partial \dot{v}} = c, \quad \frac{\partial f}{\partial v'} = -b, \quad \frac{\partial f_j}{\partial v} = -\phi_j, \quad \frac{\partial f_1}{\partial v''} = 0,$$

$$\frac{\partial f_2}{\partial v''} = -2, \quad \frac{\partial f_3}{\partial v''} = -4, \quad j = 1, 2, 3,$$

$$\phi_1 = kF_1 A_H/(AT^{(1)}), \quad R_1 = F_1[1 - vkA_H/(AT^{(1)})] + \sigma_1 a_3,$$

$$\phi_2 = A_4 F_2 A_H/A, \quad R_2 = F_2(1 - A_4 v A_H/A) + \sigma_2 a_3, \quad \phi_3 = 0,$$

$$R_3 = \sigma_3 a_3, \quad A = s + zT^m, \quad \sigma_j = 1/3, \quad j = 1, 2, 3. \quad (3.142)$$

We receive the solution of a boundary problem (3.139) and (3.140) at first on the coordinate direction $x_1$, using Eqs. (3.141), (3.142), (3.131), and (3.132) with an index (1) above. We will substitute the last in a boundary problem (3.139) and (3.140) and for simplicity of further calculations we will introduce the right-hand side of the received equation through $h_1 = c(v^{(0)})\dot{v}^{(1)} - R_1(v^{(0)})$. Then it will be:

$$\frac{\partial^2 v^{(1)}}{\partial x_1^2} = -b\frac{\partial v^{(1)}}{\partial x_1} - v^{(1)}\phi_1 + h_1. \tag{3.143}$$

Let's apply the Laplace integral transformation [6] to the differential equation (3.143), excluding derivative on $x_1$ and replacing it with linear expression concerning the image of required function. Then we consider the functions for which the Laplace integral transformation converges absolutely. The valid part of complex number $p = \xi + i\eta, i = \sqrt{-1}$ is considered positive, that is Re $p > 0$. Let's introduce capital letters for images $V, H_1$ and omit an index (1) above. We will have:

$$V(p,t,x_2,x_3) = \int_0^\infty \exp(-px_1)v(t,x)\,dx_1, v(t,x) = L^{-1}[V(p,t,x_2,x_3)],$$
$$h_1(t,x) = L^{-1}[H_1(p,t,x_2,x_3)].$$

We assume that the desired solution $v(t,x)$, and also its derivatives entering into the Eq. (3.143) satisfy Laplace integral transformation conditions on $x$, moreover growth degree on $x_1$ functions $v(t,x)$ and its derivatives do not depend on $x_2$, $x_3$. Multiplying both parts of the Eq. (3.143) on $\exp(-px_1)$ and integrating on $x_1$ from 0 to $\infty$, we will receive [6] at $(\partial v^{(1)}/\partial x_1)|_{x_1=0} = \partial g_1/\partial x_1, g_1 = v^{(1)}(t,0,x_2,x_3)$:

$$\begin{aligned} p^2V(t,p,x_2,x_3) - pg_1(t,0,x_2,x_3) - \frac{\partial g_1(t,0,x_2,x_3)}{\partial x_1} \\ + b[pV(t,p,x_2,x_3) - g_1(t,0,x_2,x_3)] + \phi_1 V(t,p,x_2,x_3) \\ = H_1(t,p,x_2,x_3), \quad 0 < x_j < L_j, \quad j = 2,3 \end{aligned}$$

$$V = \frac{(p+\delta)g_1}{(p+\delta)^2 + b_1^2} + \frac{b_1\left(\delta g_1 + \frac{\partial g_1}{\partial x_1} + H_1\right)}{b_1[(p+\delta)^2 + b_1^2]}, \tag{3.144}$$

where $\delta = b/2, b_1 = \sqrt{\phi_1 - \delta^2}$.

Using the Laplace integral transformation [6]: $L^{-1}[p/(p^2 + b_1^2)] = \cos(b_1x_1)$ at $b_1^2 > 0, L^{-1}[p/(p^2 - b_1^2)] = ch(b_1x_1)$ at $b_1^2 < 0; L^{-1}[(p + \delta)^{-1}] = \exp(-\delta x_1)$, $L^{-1}[H_1(p)/p] = \int_0^{x_1} h_1(y)\,dy$, we restore the original for $v(t,x)$ from Eq. (3.144) omitting for brevity the arguments $x_2, x_3$:

$$v(t,x) = \exp(-\delta x_1)\left\{g_1[u_1(x_1) + \delta u_2(x_1)] + u_2(x_1)\frac{\partial g_1}{\partial x_1}\right.$$
$$\left. + \int_0^{x_1} \exp(\delta y) u_2(x_1 - y) h_1(y)\, dy\right\}, \quad 0 < x_j < L_j, \quad j = 2, 3, \tag{3.145}$$

where $u_1(x_1) = \cos(b_1 x_1), u_2(x_1) = b_1^{-1}\sin(b_1 x_1), u_2(x_1 - y) = b_1^{-1}\sin[b_1(x_1 - y)]$ at $b_1^2 = \phi_1 - \delta^2 > 0$ from Eq. (3.144); $u_1(x_1) = ch(b_1 x_1), u_2(x_1) = b_1^{-1} sh(b_1 x_1), (x_1 - y) = b_1^{-1} sh[b_1(x_1 - y)]$ at $b_1^2 < 0$.

To finding an unknown derivative $\frac{\partial g_1}{\partial x_1}$ and functions $g_1$ in Eq. (3.145) we differentiate the last on $x_1$ [it is supposed that there are limited partial derivatives on $x_m, m = 1, 2, 3$ from $v(t,x)$]. Then we will receive at $b_1$ and $\delta$, obviously independent on $x_1$ omitting for simplicity of calculations the arguments at $v$:

$$\frac{\partial v^{(1)}}{\partial x_1} = \exp(-\delta x_1)\left\{\frac{\partial g_1}{\partial x_1}[u_1(x_1) - \delta u_2(x_1)]\right.$$
$$\left. - g_1 u_2(x_1)(\delta^2 \pm b_1^2) + \int_0^{x_1} \exp(\delta y) r(x_1 - y) h_1\, dy\right\}, \tag{3.146}$$

where $r(x_1 - y) = u_1(x_1 - y) - \delta u_2(x_1 - y)$.

Signs $\pm$ before $b_1^2$ are chosen in the same manner as in Eq. (3.145): before $b_1^2$ the sign "plus" is taken if $b_1^2 = \phi_1 - \delta^2 > 0$ from Eq. (3.144) or a sign "minus", if $b_1^2 < 0$.

Derivative $\frac{\partial g_1}{\partial x_1}$ and function $g_1$ in Eq. (3.146) we find using boundary conditions on the first coordinate direction $x_1$ from Eq. (3.132):

$$\left.\frac{\partial v^{(1)}}{\partial x_1}\right|_{x_1=0} = -\gamma_0(T^{(1)}|_{x_1=0} - \theta_0)/A_H = \frac{\partial g_1}{\partial x_1},$$
$$\left.\frac{\partial v^{(1)}}{\partial x_1}\right|_{x_1=L_1} = -\gamma_{L_1}(T^{(1)}|_{x_1=L_1} - \theta_{L_1})/A_H$$
$$= \exp(-\delta L_1)\left\{\frac{\partial g_1}{\partial x_1}[u_1(L_1) - \delta u_2(L_1)] - g_1 u_2(L_1)(\delta^2 \pm b_1^2)\right.$$
$$\left. + \int_0^{L_1} \exp(\delta y) r(L_1 - y) h_1\, dy\right\}. \tag{3.147}$$

Therefore finding $\frac{\partial g_1}{\partial x_1}, g_1$ in Eq. (3.147) and substituting them in Eq. (3.145), we will receive for $v$:

$$v(t,x) = \exp(-\delta x_1)\left\{P\left[B_1 + \int_0^{L_1} \exp(\delta y) r(L_1 - y) h_1\, dy\right]\right.$$
$$\left. -u_2(x_1)\gamma_0(T^{(1)}|_{x_1=0} - \theta_0)/A_H + \int_0^{x_1} \exp(\delta y) u_2(x_1 - y) h_1\, dy\right\}, \tag{3.148}$$

$$P = \frac{[u_1(x_1) + \delta u_2(x_1)]}{u_2(L_1)(\delta^2 \pm b_1^2)}, \quad B_1 = \exp(\delta L_1)\gamma_{L_1}(T^{(1)}|_{x_1=L_1} - \theta_{L_1})/A_H$$
$$- [u_1(L_1) - \delta u_2(L_1)]\gamma_0(T^{(1)}|_{x_1=0} - \theta_0)/A_H.$$

Let's transform the expression on the right-hand side (3.148) to get rid of the second integral with a variable top limit. Then, entering Green's function $E(x_1, y)$ [44]

$$E(x_1,y) = \begin{cases} \exp[\delta(y - x_1)][u_2(x_1 - y) + Pr(L_1 - y)], & 0 \le y \le x_1; \\ P\exp[\delta(y - x_1)]r(L_1 - y), & x_1 \le y \le L_1, \end{cases}$$

expression (3.148) using formulas (3.142), it will be rewritten:

$$\dot{v}^{(1)} - U_1 v^{(1)} = Z_1^{-1}\left\{\exp(-\delta x_1)[u_2(x_1)\gamma_0(T^{(1)}|_{x_1=0} - \theta_0)/A_H - PB_1]\right.$$
$$\left. + \int_0^{L_1} E(x_1,y) R_1(v^{(0)})\, dy\right\} = W_1(v^{(0)}, x, t),$$
$$U_1 = Z_1^{-1}, \quad Z_1 = \int_0^{L_1} c(v^{(0)}) E(x_1, y)\, dy,$$
$$v^{(1)}(0,x) = v_H(x), \quad 0 < x_j < L_j, \quad j = 2, 3. \tag{3.149}$$

As a result the problem solution (3.149) will be [26] [$t = t_*$ from Eq. (3.92)]:

$$v^{(1)}(x,t_*) = \left\{v_H^{(1)} + \int_0^{t_*} W_1(v^{(0)}, x, \tau)\exp[-\tau U_1(v^{(0)})]\, d\tau\right\}$$
$$\times \exp[t_* U_1(v^{(0)})], \ v_H^{(1)} = v_H(x), \ v^{(0)} = v_n, \ n = 0, 1, 2, \ldots, \tag{3.150}$$

$$T^{(1)} = T^{(0)} - f(T^{(0)})/f_T'(T^{(0)}), \quad f_T'(T^{(0)}) = \frac{\partial f(T^{(0)})}{\partial T^{(0)}},$$
$$f(T^{(0)}) = z(T^{(0)})^{m+1} + s(m+1)T^{(0)} - A_H(m+1)v^{(1)}(t_*, x),$$

$$f'_T(T^{(0)}) = z(m+1)(T^{(0)})^m + s(m+1),$$

and $0 < x_j < L_j, j = 2, 3$ are changed parametrically.

In solution of a boundary problem (3.129) is the same at Eqs. (3.137) and (3.138) or Eqs. (3.139) and (3.140) and other equations from Eqs. (3.133)–(3.136) and formulas (3.141) and (3.142) in coordinate directions $x_2, x_3$. Then similarly to Eqs. (3.144) and (3.145) we have:

$$\frac{\partial^2 v^{(2)}}{\partial x_2^2} = -b\frac{\partial v^{(2)}}{\partial x_2} - v^{(2)}\phi_2 - 2w\frac{\partial^2 v^{(2)}}{\partial x_1 \partial x_2} + h_2,$$

$$h_2 = c(v^{(1)})\dot{v}^{(2)} - R_2(v^{(1)}, x, t),$$

$$v^{(2)}(x,t) = \exp(-\delta x_2)\left\{g_2[u_1(x_2) + \delta u_2(x_2)] + u_2(x_2)\left(\frac{\partial g_2}{\partial x_2} + 2\frac{\partial g_2}{\partial x_1}\right) + \int_0^{x_2} \exp(\delta y)u_2(x_2 - y)h_2(y)\,dy - 2w\int_0^{x_2} \exp(\delta y)r(x_2 - y)\frac{\partial v^{(1)}}{\partial x_1}\,dy\right\}. \quad (3.151)$$

Value $dg_2/dx_2$ in (3.151) we will find using the second boundary condition on coordinate direction $x_2$ from Eq. (3.134):

$$\frac{\partial g_2}{\partial x_2} = u_2^{-1}(L_2)\left\{q_2\exp(\delta L_2) - g_2[u_1(L_2) + \delta u_2(L_2)] - \int_0^{L_2} \exp(\delta y)u_2(L_2 - y)h_2(y)\,dy + 2w\int_0^{L_2} \exp(\delta y)r(L_2 - y) \times \frac{\partial v^{(1)}}{\partial x_1}\,dy\right\} - 2\frac{\partial g_2}{\partial x_1}. \quad (3.152)$$

Let's enter Green's functions $Y$ and $G_2$ to get rid of integrals with a variable top limit in Eq. (3.151), then substituting in Eq. (3.151) size $\frac{\partial g_2}{\partial x_2}$ from Eq. (3.152), we will have:

$$v^{(2)}(x,t) = -S(x_2) + \int_0^{L_2} Y(x_2, y)h_2\,dy + 2w\int_0^{L_2} G_2(x_2, y)\frac{\partial v^{(1)}}{\partial x_1}\,dy,$$

$$S(x_2) = \exp(-\delta x_2)u_2(x_2)\{[u_1(L_2) + \delta u_2(L_2)]g_2 - q_2\exp(\delta L_2)\}/u_2(L_2) - \exp(-\delta x_2)[u_1(x_2) + \delta u_2(x_2)]g_2; \quad (3.153)$$

$$Y(x_2, y) = \begin{cases} \frac{\exp[\delta(y-x_2)]}{u_2(L_2)}[u_2(x_2 - y)u_2(L_2) - u_2(x_2)u_2(L_2 - y)], \\ \quad 0 \le y \le x_2; \\ -\exp[\delta(y - x_2)]u_2(x_2)u_2(L_2 - y)/u_2(L_2), \quad x_2 \le y \le L_2, \end{cases}$$

$$G_2(x_2, y) = \begin{cases} \exp[\delta(y - x_2)]\frac{\{u_2(x_2)[u_1(L_2-y)-\delta u_2(L_2-y)]}{u_2(L_2)} \\ \quad -\frac{u_2(L_2)[u_1(x_2-y)-\delta u_2(x_2-y)]\}}{u_2(L_2)}, \quad 0 \le y \le x_2; \\ \exp[\delta(y - x_2)]\frac{u_2(x_2)[u_1(L_2-y)-\delta u_2(L_2-y)]}{u_2(L_2)}, \quad x_2 \le y \le L_2. \end{cases}$$

To find $\frac{\partial v^{(1)}}{\partial x_1}$ we use formulas (3.146) and (3.147) so introducing Green's function we receive $E_1(x_1, y)$:

$$\frac{\partial v^{(1)}}{\partial x_1} = \exp(-\delta x_1)\{[u_1(x_1) - \delta u_2(x_1)]\gamma_0(\theta_0 - T^{(1)}|_{x_1=0})/A_H$$
$$- \frac{u_2(x_1)}{u_2(L_1)}B_1 + \int_0^{L_1} E_1(x_1, y)h_1\, dy; \tag{3.154}$$

$$E_1(x_1, y) = \begin{cases} \exp[\delta(y - x_1)]\frac{[u_2(L_1)r(x_1-y)-u_2(x_1)r(L_1-y)]}{u_2(L_1)}, \quad 0 \le y \le x_1; \\ -\exp[\delta(y - x_1)]\frac{u_2(x_1)r(L_1-y)}{u_2(L_1)}, \quad x_1 \le y \le L_1. \end{cases}$$

As a result in the second coordinate direction $x_2$ we receive the formula of a type (3.149)

$$\dot{v}^{(2)} - U_2 v^{(2)} = Z_2^{-1}\left[S(x_2) + \int_0^{L_2} Y(x_2, y)R_2(v^{(1)})\, dy \right.$$
$$\left. -2w\int_0^{L_2} G_2(x_2, y)\frac{\partial v^{(1)}}{\partial x_1}\, dy\right]$$
$$= W_2(v^{(1)}, x, t), \quad Z_2 = \int_0^{L_2} c(v^{(1)})Y(x_2, y)\, dy, \quad U_2 = Z_2^{-1};$$
$$v^{(2)}(0, x) = v^{(1)}(t_*, x), \quad 0 \le x_1 \le L_1, \quad 0 < x_3 < L_3. \tag{3.155}$$

As a result the problem solution (3.155) on the coordinate direction $x_2$ looks like

$$v^{(2)}(x, t_*) = \left\{v_H^{(2)} + \int_0^{t_*} W_2(v^{(1)}, x, \tau)\exp[-\tau U_2(v^{(1)})]\, d\tau\right\}$$
$$\times \exp[t_* U_2(v^{(1)})], \quad v_H^{(2)} = v_1^{(1)}(t_*, x), \tag{3.156}$$

where $T^{(2)}$ is received from Eq. (3.150) by replacement of the top indexes (1) and (0), accordingly, on (2) and (1).

Last, in the third coordinate direction $x_3$

$$\frac{\partial^2 v^{(3)}}{\partial x_3^2} = -b\frac{\partial v^{(3)}}{\partial x_3} - v^{(3)}\phi_3 - 2w\left(\frac{\partial^2 v^{(3)}}{\partial x_1 \partial x_3} + \frac{\partial^2 v^{(3)}}{\partial x_2 \partial x_3}\right) + h_3,$$

$$h_3 = c(v^{(2)})\dot{v}^{(3)} - R_3(v^{(2)}, x, t),$$

using the algorithm (3.144), (3.145), (3.151)–(3.153), we will find:

$$v^{(3)}(t,x) = -S(x_3) + \int_0^{L_3} Y(x_3, y)h_3\, dy$$
$$+ 2w\int_0^{L_3} G_2(x_3, y)\left(\frac{\partial v^{(1)}}{\partial x_1} + \frac{\partial v^{(2)}}{\partial x_2}\right) dy. \tag{3.157}$$

To find a derivative $\frac{\partial v^{(2)}}{\partial x_2}$ in Eq. (3.157) we differentiate the expression (3.151) on $x_2$ so we have:

$$\frac{\partial v^{(2)}}{\partial x_2} = -2w\frac{\partial v^{(1)}}{\partial x_1} + \exp(-\delta x_2)\left\{\left(\frac{\partial g_2}{\partial x_2} + 2\frac{\partial g_2}{\partial x_1}\right)[u_1(x_2) - \delta u_2(x_2)]\right.$$
$$- g_2 u_2(x_2)(\delta^2 \pm b_2^2) + \int_0^{x_2} \exp(\delta y) r(x_2 - y) h_2\, dy$$
$$\left. +2w\int_0^{x_2} \exp(\delta y) o(x_2 - y)\frac{\partial v^{(1)}}{\partial x_1}\, dy\right\}, \tag{3.158}$$

where $o(x_2 - y) = \delta r(x_2 - y) + \delta u_1(x_2 - y) \pm b_2^2 u_2(x_2 - y)$.

Taking $\frac{\partial g_2}{\partial x_2}$ from Eq. (3.152) and repeatedly introducing Green's functions $G_1, G_3$, we will get rid of integrals with a variable top limit in Eq. (3.158). Finally (3.158) we have:

$$\frac{\partial v^{(2)}}{\partial x_2} = V(x_2) + \int_0^{L_2} G_1(x_2, y)h_2\, dy - 2\frac{\partial v^{(1)}}{\partial x_1} + 2\int_0^{L_2} G_3(x_2, y)\frac{\partial v^{(1)}}{\partial x_1}\, dy; \tag{3.159}$$

$$V(x_2) = \exp(-\delta x_2)\frac{\{q_2 \exp(\delta L_2) - g_2[u_1(L_2) + \delta u_2(L_2)]\}}{u_2(L_2)}$$
$$\times [u_1(x_2) - \delta u_2(x_2)] - \exp(-\delta x_2) g_2 u_2(x_2)(\delta^2 \pm b_2^2),$$

$$G_1(x_2, y) = \begin{cases} \exp[\delta(y - x_2)]\frac{\{u_2(L_2)r(x_2-y) - [u_1(x_2)}{u_2(L_2)} - \frac{\delta u_2(x_2)]u_2(L_2-y)\}}{u_2(L_2)}, \\ \quad 0 \le y \le x_2; \\ -\exp[\delta(y - x_2)]\frac{[u_1(x_2) - \delta u_2(x_2)]u_2(L_2-y)}{u_2(L_2)}, \quad x_2 \le y \le L_2, \end{cases}$$

$$G_3(x_2,\gamma) = \begin{cases} \exp[\delta(\gamma - x_2)]\frac{\{[u_1(x_2)-\delta u_2(x_2)][u_1(L_2-\gamma)}{u_2(L_2)} \\ \quad + \frac{-\delta u_2(L_2-\gamma)]+u_2(L_2)o(x_2-\gamma)\}}{u_2(L_2)}, \quad 0 \le \gamma \le x_2; \\ \exp[\delta(\gamma - x_2)]\frac{[u_1(x_2)-\delta u_2(x_2)][u_1(L_2-\gamma)}{u_2(L_2)} - \frac{\delta u_2(L_2-\gamma)]}{u_2(L_2)}, \\ x_2 \le \gamma \le L_2. \end{cases}$$

As a result in the third coordinate direction we find the formula similar to Eq. (3.155):

$$\dot{v}^{(3)} - U_3 v^{(3)} = Z_3^{-1}\left[S(x_3) + \int_0^{L_3} Y(x_3,\gamma)R_3(v^{(2)})\,d\gamma \right.$$
$$\left. -2\int_0^{L_3} G_2(x_3,\gamma)\left(\frac{\partial v^{(1)}}{\partial x_1} + \frac{\partial v^{(2)}}{\partial x_2}\right)d\gamma\right] = W_3(v^{(2)},x,t),$$
$$v^{(3)}(0,x) = v^{(2)}(t_*,x), \quad 0 \le x_1 \le L_1, \quad 0 < x_2 < L_2,$$
$$Z_3 = \int_0^{L_3} c(v^{(2)})Y(x_3,\gamma)\,d\gamma, \quad U_3 = Z_3^{-1}, \tag{3.160}$$

where $\frac{\partial v^{(1)}}{\partial x_1}, \frac{\partial v^{(2)}}{\partial x_2}$ are defined according to formulas (3.154) and (3.159). As a result the final problem solution (3.160) will be:

$$v^{(3)}(x,t_*) = \{v_H^{(3)} + \int_0^{t_*} W_3(v^{(2)},x,\tau)\exp[-\tau U_3(v^{(2)})]\,d\tau$$
$$\times \exp[t_* U_3(v^{(2)})], \quad v_H^{(3)} = v^{(2)}(t_*,x),$$
$$0 < x_2 < L_2, \quad 0 \le x_1 \le L_1, \quad n = 0,1,2\ldots. \tag{3.161}$$

To receive $S(x_3)$ in Eq. (3.160) it is necessary in the formula $S(x_2)$ from relationship (3.153) to replace the arguments $x_2, L_2$ according to $x_3, L_3$. Thus $T^{(3)}$ is received from Eq. (3.150) by replacement of the top indexes (1) and (0) accordingly on (3) and (2).

At $x = x_2$ in Eq. (3.156) other variables of $0 \le x_1 \le L_1, 0 < x_3 < L_3$ are changed parametrically, as in Eq. (3.150). A similar situation is for $x = x_3$ in Eq. (3.161), thus the final solution of the first boundary problem is (3.129): $v^{(3)}(t_*,x) = v_{n+1}(t_*,x), \forall t_* > 0, n = 0,1,2,\ldots$, and using the inversion formula (3.125)—of initial nonlinear boundary problem (3.118)–(3.120).

If in Eq. (3.134) in a coordinate direction $x_2$ the boundary condition of the 3rd type is introduced

$$\left.\frac{\partial v^{(2)}}{\partial x_2}\right|_{x_2=0} = -\gamma_0(T^{(2)}|_{x_2=0} - \theta_0)/A_H,$$

$$\left.\frac{\partial v^{(2)}}{\partial x_2}\right|_{x_2=L_2} = -\gamma_{L_2}(T^{(2)}|_{x_2=L_2} - \theta_{L_2})/A_H,$$

using formulas (3.146) and (3.147) for $\frac{\partial v^{(2)}}{\partial x_2}$ instead of the expression (3.159) we find:

$$\begin{aligned}\frac{\partial v^{(2)}}{\partial x_2} &= \int_0^{L_2} E_2(x_2, y)h_2\, dy - 2w\frac{\partial v^{(1)}}{\partial x_1} + 2w\int_0^{L_2} E_3(x_2, y)\frac{\partial v^{(1)}}{\partial x_1}\, dy \\ &+ \exp[\delta(L_2 - x_2)]B_2\left[2w\left.\frac{\partial v^{(1)}}{\partial x_1}\right|_{x_2=L_2} - \gamma_{L_2}(T^{(2)}|_{x_2=L_2} - \theta_{L_2})/A_H\right] \\ &+ \exp(-\delta x_2)\left[2w\left.\frac{\partial v^{(1)}}{\partial x_1}\right|_{x_2=0} - \gamma_0(T^{(2)}|_{x_2=0} - \theta_0)/A_H\right] \\ &\times \{u_1(x_2) - \delta u_2(x_2) - B_2[u_1(L_2) - \delta u_2(L_2)]\},\end{aligned}$$

$$B_2 = \frac{u_2(x_2)(\delta^2 \pm b_2^2)}{u_2(L_2)(\delta^2 \pm b_2^2)|_{x_2=L_2}},$$

$$E_2(x_2, y) = \begin{cases} \frac{\exp[\delta(y-x_2)]}{u_2(L_2)}[r(x_2 - y)u_2(L_2) - u_2(x_2)r(L_2 - y)], & 0 \le y \le x_2; \\ -\exp[\delta(y - x_2)]u_2(x_2)r(L_2 - y)/u_2(L_2), & x_2 \le y \le L_2, \end{cases}$$

$$E_3(x_2, y) = \begin{cases} \frac{\exp[\delta(y-x_2)]}{u_2(L_2)}[o(x_2 - y)u_2(L_2) - u_2(x_2)o(L_2 - y)], & 0 \le y \le x_2; \\ -\exp[\delta(y - x_2)]u_2(x_2)o(L_2 - y)/u_2(L_2), & x_2 \le y \le L_2. \end{cases}$$

If in Eq. (3.132) in a coordinate direction $x_1$ the boundary condition of the 1st type is introduced

$$v^{(1)}|_{x_1=0} = g_1(t, x_2, x_3), \quad v^{(1)}|_{x_1=L_1} = q_1(t, L_1, x_2, x_3),$$

according to the algorithm (3.152), (3.158), and (3.159) we find:

$$\frac{\partial v^{(1)}}{\partial x_1} = V(x_1) + \int_0^{L_1} G_1(x_1, y)h_1\, dy,$$

where $V(x_1)$ is received from $V(x_2)$ in Eq. (3.159) by replacement of arguments $x_2, L_2$ accordingly on $x_1, L_1$ and the bottom index 2 at functions $g_2, q_2, b_2$ on an index 1.

Using the results of Chapter 2, it is easy to find the conditions of unequivocal resolvability of a boundary problem (3.118), (3.119), and (3.121) in the absence of the mixed derivatives and $B = 0, A_2 = A_3 = 0,$

$C = A_1 = 1$ and also to receive a quadratic estimation of speed of convergence of the iterative process.

## Results of Test Checks

Accuracy of the received approximate analytical formulas (3.150), (3.156), and (3.161) we will check practically in solution of the boundary problem for the equation in partial derivatives. For simplicity of the analysis first of all we will consider boundary conditions of the 1st type in the region $Q : (0 \le x_j \le b, j = 1, 2, 3), 0 \le t \le t_k$:

$$A_6 \frac{\partial T}{\partial t} = \sum_{j=1}^{3} \frac{\partial}{\partial x_j}\left(A_2 \frac{\partial T}{\partial x_j}\right) + A_1 \sum_{j=1}^{3} \frac{\partial T}{\partial x_j} + F(x, t)$$

$$+ w \sum_{j=1, i\neq j}^{3} \frac{\partial}{\partial x_j}\left(A_2 \frac{\partial T}{\partial x_i}\right) + A_3 T^k + A_4 \exp(A_5 T); \quad (3.162)$$

$$T|_{t=0} = \exp(y_1 + y_2 + y_3), \quad y_j = x_j/b, \quad (3.163)$$

$$T|_{x_1=0} = \exp(t + y_2 + y_3), \quad T|_{x_2=0} = \exp(t + y_1 + y_3),$$

$$T|_{x_3=0} = \exp(t + y_1 + y_2), \quad T|_{x_j=b} = \exp(1)$$

$$\times T|_{x_j=0}, \quad j = 1, 2, 3, \quad (3.164)$$

and then boundary conditions of the mixed type:

$$\left.\frac{\partial T}{\partial x_1}\right|_{x_1=0} = -\gamma_0 (T|_{x_1=0} - \theta_0)/A_H,$$

$$\left.\frac{\partial T}{\partial x_1}\right|_{x_1=b} = -\gamma_{L_1} (T|_{x_1=b} - \theta_{L_1})/A_H;$$

$$T|_{x_2=0} = \exp(t + y_1 + y_3),$$

$$T|_{x_3=0} = \exp(t + y_1 + y_2),$$

$$T|_{x_j=b} = \exp(1) \cdot T|_{x_j=0}, \quad j = 2, 3. \quad (3.165)$$

The exact solution of a problem (3.162)–(3.164) was taken $T = \exp(t + u), u = \sum_{j=1}^{3} y_j$, then a source $F$ in the Eq. (3.162) will be:

$$F = \exp(t + u)\{A_6 - 3A_1/b - 3[s + z(m + 1)\exp(m(t + u))](1 + 2w)/b^2\}$$
$$- A_3 \exp[k(t + u)] - A_4 \exp[A_5 \exp(t + u)],$$
$$A_2 = s + zT^m, \quad m \neq -1.$$

The boundary problem (3.162)–(3.164) is solved with the help of formulas of types (3.156) and (3.161). The program is made in the language Fortran-90, calculation was made on Pentium 3 (800 MHz, the compiler PS 5) with double accuracy. The maximum relative was fixed error

$$\epsilon = \frac{|T - \tilde{T}|100\%}{T},$$

where $T$ is exact, $\tilde{T}$ is approximate analytical solution. The number of iterations ($J$) was traced [for total expressions of a type (3.161) according to formulas (3.49)–(3.53)] on relative change of an error vector:

$$||V_n|| = \max_{x,t \in Q} \left| \frac{v_{n+1} - v_n}{v_{n+1}} \right| .$$

Results of test calculations are given for $||V_n|| \leq \delta, \delta = 0.01$. Thus calculation time of any variant is $t_p = 5$ s.

The following initial data was used: $N_j = 11, \Delta x_j = b/(N_j - 1), j = 1, 2, 3, M = 101, \Delta t = t/(M - 1); N_j, M, \Delta x_j, j = 1, 2, 3, \Delta t$ is a number of checkouts and steps on space and time while finding integrals in the Eqs. (3.149) and (3.150) by Simpson's formula [21]. Results are received for the following basic values of initial data: $N_j = 11, j = 1, 2, 3, M = 11, m = 0.5, s = A_H = 1, t = 1, A_5 = -1$. In Table 3.4 there are results of calculations of a boundary problem (3.162)–(3.164) for $A_6 = 10^6, z = 0.01$ at various values $b, w, k, A_1, A_3, A_4$.

In modeling convective-conductive heat-transfer in a problem of thermal protection [31] materials with high $A_6$ (carbon-plastics, asbocement, and others $A_6 : 10^6–4 \cdot 10^6$ s/m$^2$), and with low $A_6$ (copper:

**Table 3.4** A dependence of the maximum relative errors at the various values $b, w, k, A_1, A_3, A_4$

| | Results of calculations | | | | | | | |
|---|---|---|---|---|---|---|---|---|
| Variant number | w | k | $A_1$ | $A_3$ | $A_4$ | b | ε, % | J |
| 1 | 0 | 1 | 1 | 0 | 0 | 0.01 | 0.138 | 2 |
| 2 | 0 | 1 | 1 | 0 | 0 | 0.05 | 0.173 | 2 |
| 3 | 0 | 1 | 1 | 0 | 0 | 0.1 | 0.209 | 2 |
| 4 | 0.5 | 1 | 1 | 1 | 0 | 0.1 | 0.183 | 2 |
| 5 | 0.5 | 2 | 1 | 1 | 0 | 0.1 | 0.202 | 2 |
| 6 | 0.5 | 1 | 1 | 0 | 1 | 0.1 | 0.184 | 2 |
| 7 | 0.5 | 1 | 100 | 0 | 0 | 0.1 | 0.278 | 2 |
| 8 | 0.5 | 1 | −20 | 0 | 0 | 0.1 | 0.165 | 2 |
| 9 | 0.5 | 1 | 1 | 500 | 0 | 0.1 | 0.214 | 2 |
| 10 | 0.5 | 1 | 1 | −500 | 0 | 0.1 | 0.171 | 2 |

$A_6 = 9 \cdot 10^3\ \mathrm{s/m^2}$) are interesting. Besides, the geometrical sizes of area of problem definition (thickness of a heat-shielding covering) are less than 1 m. Results presented in Table 3.5 are calculations of the mixed boundary problem (3.162), (3.163), and (3.165) by formulas (3.150), (3.156), and (3.161) at $z = \gamma_0 = \gamma_{L_1} = 1, b = 0.05, A_6 = 9 \cdot 10^3, \theta_{L_1} = \theta_0 = w = 0$ for a basic variant. At $w = 0.5$ and a type of carbon-plastics materials ($A_6 = 10^6, z = 0.01$) for a variant at number 1 from Table 3.5 we used $\epsilon = 0.264 (J = 2)$. Here, it is necessary to notice that the given technology allows to solve the stationary ($A_6 = 0$) equation (3.162). In this case under boundary conditions (3.163), and (3.164), independent on time, for a variant at number 3 from Table 3.5 we have $\epsilon = 0.0147 (J = 2)$. At the same time numerical realization of stationary boundary problem (3.162)–(3.164) is based on iterative methods [12] with relaxation parameter $\omega (0 < \omega < 2)$, or on an establishment method (at $t \to \infty$) [23] for the corresponding non-stationary equation with stationary (not dependent on time) boundary conditions. These numerical approaches lead to considerable expense of machine time especially in three-dimensional space.

Besides, for iterative methods [23] at $A_6 = 0$ and boundary conditions of the 2nd sort $\frac{\partial T}{\partial n}|_\Gamma = q (q = 0, q \neq 0)$ numerical realization becomes complicated on implicit difference to the scheme on space, if the Eq. (3.118) is used for reception of a high error of approximation and unconditional stability.

In this case the first proracing coefficients are equal $-1$ [even at $B = 0$ in Eq. (3.118)]. It leads to the necessity to use formulas of nonmonotonic prorace [23] that are undesirable for known in advance monotonous solutions of an initial boundary problem. At the same time

**Table 3.5** A dependence of the maximum relative errors at the various values $k, A_1, A_3, A_4$

| | Results of calculations | | | | | |
|---|---|---|---|---|---|---|
| **Variant number** | $A_1$ | $A_3$ | $A_4$ | $k$ | $\varepsilon$, % | $J$ |
| 1 | 1 | 1 | 0 | 1 | 0.179 | 2 |
| 2 | 1 | −1 | 0 | 1 | 0.179 | 2 |
| 3 | 1 | −1 | 0 | 2 | 0.178 | 2 |
| 4 | 40 | 1 | 0 | 1 | 0.184 | 2 |
| 5 | −100 | 1 | 0 | 1 | 0.172 | 2 |
| 6 | 1 | 500 | 0 | 1 | 0.183 | 2 |
| 7 | 1 | −5000 | 0 | 1 | 0.136 | 2 |
| 8 | 1 | 1 | 1 | 1 | 0.179 | 2 |

on technology (3.150), (3.156), and (3.161) $b = 0.1, z = 0.01, A_1 = -1$, $k = A_3 = 1, \gamma_0 = \gamma_{L_1} = w = A_6 = 0$ in solution of a boundary problem (3.162), (3.163), and (3.165) with stationary (not dependent on time) boundary conditions, we receive $\epsilon = 0.0116 (J = 2)$.

## CONCLUSION

Thus, using the inversion formula (3.125) in the wide range of change of initial data given $10^{-2} \leq b \leq 0.1, 9 \cdot 10^3 \leq A_6 \leq 10^6, A_6 = 0, |A_1| \leq 100, -5000 \leq A_3 \leq 500$ we receive the estimation of approximate analytical solution of the modeling of nonlinear boundary problems (3.162)–(3.164) and (3.162), (3.163), and (3.165)

By means of methods quasi-linearization, locally-one-dimensional scheme splittings, Laplace integral transformation and on a concrete example, it is shown that the given technology by efficiency does not concede to the numerical solution.

# CHAPTER 4

# Method of Solution of Conjugate Boundary Problems

In this chapter, which covers conjugate problems of heat-exchange, approximate analytical solutions of nonlinear boundary problems in one-dimensional and spatial statements are provided. For a nonlinear one-dimensional and three-dimensional boundary problem, the estimation of speed of convergence of the iterative process is found. Comparison of accuracy of solution of one one-dimensional problem of heat-exchange [45] on offered mathematical technology with the known analytical solution of this problem is shown [45]. In a spatial case by a method of trial functions, we show the results of test checks of mathematical technology and a comparison with the known numerical methods and analytical solutions are discussed.

## 4.1 METHOD OF SOLUTION OF CONJUGATE BOUNDARY PROBLEMS

### Statement of a Problem and a Method Algorithm

Let us attempt to find a solution of the equation of a parabolic type [5, 7] adjoining bodies (layered environments) when heat-exchange between them occurs with the law of conservation of energy and in statement of a corresponding boundary problem, boundary conditions of the fourth general type take place (balance relations [2, 45]) with sources

$$C_i(T_i)\frac{\partial T_i}{\partial t} = \frac{\partial}{\partial z}\left[A_i(T_i)\frac{\partial T_i}{\partial z}\right] + B_i(T_i)\frac{\partial T_i}{\partial z} + Y_i T_i^k + E_i(z,t), \quad i = 1,2 \tag{4.1}$$

at $i = 1$ in the region $Q_1 = (a_1 < z < 0, -\infty < a_1 < 0)$ at $i = 2$ in the region $Q_2 = (0 < z < a_2, 0 < a_2 < \infty)$, $\overline{Q_i} = Q_i + \Gamma$, $\overline{Q_{t,i}} = \overline{Q_i} \times [0 < t \le t_0], i = 1,2$ for $Y_j = \text{const}, j = 1,2$ with the initial condition

$$T_i \mid_{t=0} = T_{H,i}(z), \quad i = 1,2, \tag{4.2}$$


http://dx.doi.org/10.1016/B978-0-12-804289-2.00004-1

with a balance boundary condition [2, 45]:

$$A_1 \frac{\partial T_1}{\partial z}\bigg|_{z=-0} + F(t,-0) = A_2 \frac{\partial T_2}{\partial z}\bigg|_{z=+0}, \qquad T_1|_{z=-0} = T_2|_{z=+0} \quad (4.3)$$

and on border $\Gamma$ with boundary conditions of the second, the third types

$$\left[A_1 \frac{\partial T_1}{\partial z} + S_1(t)T_1\right]_{z=-a_1} = D_1(-a_1,t),$$
$$\left[A_2 \frac{\partial T_2}{\partial z} + S_2(t)T_2\right]_{z=a_2} = D_2(a_2,t) \quad (4.4)$$

at $S_1 = S_2 = 0$ Neumann's condition and a boundary Dirichlet condition are considered below.

Let's assume everywhere:

1. A problem (4.1)–(4.4) has the unique solution $T_i(z,t), i = 1,2$ which is continuous in each region $\overline{Q_{t,i}}, i = 1,2$ and has continuous derivatives $\frac{\partial T_i}{\partial t}, \frac{\partial T_i}{\partial z}, \frac{\partial^2 T_i}{\partial z^2}, i = 1,2.$
2. The following conditions are provided: $A_i(T_i) \geq c_1 > 0, C_i(T_i) \geq c_2 > 0$, $c_1, c_2$ are constants, $i = 1,2$; $T_{H,i}$ is a continuous function in $\overline{Q_i}, i = 1,2$, and $F, A_i, B_i, E_i, i = 1,2$ are continuous functions in the closed regions $\overline{Q_{t,i}}, i = 1,2$.
3. In a general case, coefficients $C_i(T_i), B_i(T_i)$ can be nonlinearly dependent on the problem solution $T_i$ [45], the type $A_i(T_i)$ is defined below in the formula (4.6), and $S_i, D_i, i = 1,2$ are continuous functions on $\Gamma$ for $0 < t \leq t_0$ having limited partial derivatives of the first order.

Let's use the Kirchhoff's transformation [5]

$$v_i = \int_0^{T_i} \frac{A_i(T_i)}{A_{H,i}}\, dT_i, \quad i = 1,2, \quad (4.5)$$

where $A_{H,i}, i = 1,2$ is for example, the heat conductivity factor at a temperature equal to zero.

Then, to use the inversion formula, $A_i(T_i)$, $i = 1,2$ in Eq. (4.1) we take

$$A_i(T_i) = N_i T_i^m, \quad m > 0, \ N_i > 0, \ N_i = \text{const}, \ i = 1,2. \quad (4.6)$$

Taking into account [5]:

$$\frac{\partial A_i}{\partial z} = \frac{\partial A_i}{\partial T_i}\frac{\partial T_i}{\partial z}, \quad \frac{\partial v_i}{\partial t} = \frac{A_i}{A_{H,i}}\frac{\partial T_i}{\partial t}, \quad \frac{\partial v_i}{\partial z} = \frac{A_i}{A_{H,i}}\frac{\partial T_i}{\partial z}, \quad i = 1,2 \quad (4.7)$$

we receive a boundary problem from Eqs. (4.1)–(4.7)

$$\frac{\partial^2 v_i}{\partial z^2} = c_i(v_i)\frac{\partial v_i}{\partial t} - \delta_i(v_i)\frac{\partial v_i}{\partial z} - \gamma_i(\phi v_i)^{k/s} - e_i(z,t), \quad i = 1,2, \tag{4.8}$$

$$v_i\,|_{t=0} = v_{H,i}(z), \quad v_{H,i} = \frac{T^s_{H,i}}{\phi_i}, \quad \phi_i = \frac{sA_{H,i}}{N_i}, \quad i = 1,2, \; s = m+1, \tag{4.9}$$

$$\eta\left.\frac{\partial v_1}{\partial z}\right|_{z=-0} + M = \left.\frac{\partial v_2}{\partial z}\right|_{z=+0}, \quad \xi v_1|_{z=-0} = v_2|_{z=+0},$$

$$\eta = \frac{A_{H,1}}{A_{H,2}}, \quad M = \frac{F(t,-0)}{A_{H,2}}, \quad \xi = \eta\frac{N_2}{N_1}, \tag{4.10}$$

$$\left[\frac{\partial v_1}{\partial z} + \frac{S_1(v\phi)_1^{1/s}}{A_{H,1}}\right]_{z=-a_1} = \frac{D_1(t,-a_1)}{A_{H,1}},$$

$$\left[\frac{\partial v_2}{\partial z} + \frac{S_2(v\phi)_2^{1/s}}{A_{H,2}}\right]_{z=a_2} = \frac{D_2(t,a_2)}{A_{H,2}}, \tag{4.11}$$

where $c_i = C_i/A_i, \delta_i = B_i/A_i, \gamma_i = Y_i/A_{H,i}, e_i = E_i/A_{H,i}, i = 1,2$. Thus $T_i, i = 1,2$ it is defined from Eq. (4.5) according to the inversion formula:

$$v_i = T_i^s/\phi_i, \quad T_i = (v\phi)_i^{1/s}, \quad i = 1,2. \tag{4.12}$$

Let's notice that variation range of independent variables and type of the boundary conditions do not change in relation to Kirchhoff's transformation (4.5) and within the inversion formula (4.12) the boundary conditions of the first-fourth types passes into Dirichlet, Neumann, Newton's condition, and the balance boundary conditions.

As $B_i(T_i), i = 1,2$ in Eq. (4.1) obviously does not depend on $z$ we take substitution: $v_i = w_i \exp(-z\delta_i/2), i = 1,2$ [37] in Eqs. (4.8)–(4.12) to exclude the first partial derivative on space $z$ in Eq. (4.8). Then a boundary problem (4.8)–(4.11) and the inversion formula (4.12) will be rewritten

$$\frac{\partial^2 w_i}{\partial z^2} = c_i(w_i)\frac{\partial w_i}{\partial t} - \exp(r_i z)\{\gamma_i[(\phi w)_i \exp(-r_i z)]^{k/s} + e_i\} + w_i r_i^2, \quad i = 1,2, \tag{4.13}$$

$$w_i\,|_{t=0} = w_{H,i}(z), \quad w_{H,i} = T^s_{H,i}\exp(r_i z)/\phi_i, \quad r_i = 0.5\delta_i, \quad i = 1,2, \tag{4.14}$$

$$\eta\left(\frac{\partial w_1}{\partial z}-r_1w_1\right)\bigg|_{z=-0}+M=\left(\frac{\partial w_2}{\partial z}-r_2w_2\right)\bigg|_{z=+0},$$
$$\xi w_1|_{z=-0}=w_2|_{z=+0}, \tag{4.15}$$

$$\left\{\frac{\partial w_1}{\partial z}+\exp(-r_1a_1)S_1(-a_1,t)[(w\phi)_1\exp(r_1a_1)]^{1/s}/A_{H,1}-r_1w_1\right\}_{z=-a_1}$$
$$=\exp(-r_1a_1)D_1(-a_1,t)/A_{H,1},$$
$$\left\{\frac{\partial w_2}{\partial z}+\exp(r_2a_2)S_2(a_2,t)[(w\phi)_2\exp(-r_2a_2)]^{1/s}/A_{H,2}-r_2w_2\right\}_{z=a_2}$$
$$=\exp(r_2a_2)D_2(a_2,t)/A_{H,2}, \tag{4.16}$$

$$w_i=T_i^s\exp(r_iz)/\phi_i,\quad T_i=[(w\phi)_i\exp(-r_iz)]^{1/s},\quad i=1,2,$$
$$T_0=(w\phi)_0^{1/s},\quad \phi_0=s(A_{H,1}+A_{H,2})/(N_1+N_2). \tag{4.17}$$

Our purpose is to have the solution to a nonlinear boundary problem if it exists as a limit of sequence of solutions of linear boundary problems. For this purpose let's take the results of works [7, 15, 16]. Let be $w_i^0 =$ const, $i = 1, 2$ be some initial approximation [as initial approximation it is better to take $w_{H,i}, i = 1, 2$ from Eq. (4.14)]. We will consider sequence $w_i^n(t, z), i = 1, 2$ defined by a recurrence relationship [7] (the point above means a partial derivative on time):

$$\frac{\partial^2 w_i^{n+1}}{\partial z^2}=f_i+(w_i^{n+1}-w_i^n)\frac{\partial f_i}{\partial w_i^n}+(\dot{w}_i^{n+1}-\dot{w}_i^n)\frac{\partial f_i}{\partial \dot{w}_i^n},$$
$$f_i=f_i(w_i^n,\dot{w}_i^n),\quad i=1,2,\quad n=0,1,2,\ldots, \tag{4.18}$$

$$\eta\frac{\partial w_1^n}{\partial z}\bigg|_{z=-0}+Ew_0^n+M=\frac{\partial w_2^n}{\partial z}\bigg|_{z=+0},\quad w_0^n=w_1^n|_{z=0},$$
$$\xi w_1^n|_{z=-0}=w_2^n|_{z=+0},\quad E=(\xi r_2-\eta r_1)|_{z=0},$$
$$w_{H,i}=w_i^n(0,z),\quad \frac{\partial w_i^{n+1}}{\partial z}\bigg|_{\Gamma}=\left[\alpha_i+(w_i^{n+1}-w_i^n)\frac{\partial\alpha_i}{\partial w_i^n}\right]\bigg|_{\Gamma},$$
$$i=1,2,\quad \alpha_i=r_iw_i^n+\frac{1}{A_{H,i}}\exp(r_iz)\{D_i-S_i[w_i^n\phi_i$$
$$\times\exp(-zr_i)]^{1/s}\},\quad i=1,2,\quad n=0,1,2,\ldots, \tag{4.19}$$

where $i = 1$ at $z = -a_1$, $i = 2$ at $z = a_2$, and the type $f_i, i = 1, 2$ will be defined below at the concrete solution of a boundary problem (4.13)–(4.16).

Each function $w_i^{n+1}(t,z), i = 1,2$ in Eqs. (4.18) and (4.19) is a solution of the linear equation that is a rather important feature of this algorithm. The algorithm comes from an approximation Newton-Kantorovich's method [8] in functional space.

To reduce further records we will introduce the notations:

$$f_i = c_i\dot{w}_i^n - \exp(r_i z)[\gamma_i(Z_i^n)^{k/s} + e_i] + w_i^n r_i^2, \quad i = 1,2,$$

$$\Phi_i = \frac{\gamma_i k(Z_i^n)^{k/s} A_{H,i}}{Z_i^n N_i} - r_i^2, \quad r_i = 0.5\frac{B_i}{A_i}, \quad Z_i^n = w_i^n \exp(-r_i z)\phi_i,$$

$$R_i = \gamma_i(Z_i^n)^{k/s}\left[\exp(r_i z) - \frac{w_i^n k A_{H,i}}{Z_i^n N_i}\right] + e_i \exp(r_i z),$$

$$s_i = \frac{S_i(Z_i^n)^{1/s}}{Z_i^n N_i} - r_i, \quad q_i(z,t) = \frac{\exp(r_i z)}{A_{H,i}}[D_i(t) - (Z_i^n)^{1/s} S_i(t)]$$

$$+ \frac{w_i^n S_i(Z_i^n)^{1/s}}{Z_i^n N_i}, \quad c_i(w_i^n) = \frac{\partial f_i}{\partial \dot{w}_i^n}, \quad \Phi_i(w_i^n) = -\frac{\partial f_i}{\partial w_i^n}, \quad i = 1,2, \tag{4.20}$$

where for $g_i$ and $q_i$: $i = 1$ at $z = -a_1$, $i = 2$ at $z = a_2$.

Let's take the solution of a boundary problem (4.13)–(4.16), using for this purpose the Eqs. (4.17)–(4.20), so introducing them, we have

$$h_i = c_i\dot{w}_i^{n+1} - R_i(w_i^n), \quad i = 1,2, \tag{4.21}$$

$$\frac{\partial^2 w_i^{n+1}}{\partial z^2} + w_i^{n+1}\Phi_i = h_i(w_i^n, \dot{w}_i^{n+1}, z, t), \quad i = 1,2, \tag{4.22}$$

$$w_i^n|_{t=0} = w_{H,i}, \quad i = 1,2, \quad n = 0,1,2,\ldots, \tag{4.23}$$

$$\eta\left.\frac{\partial w_1^n}{\partial z}\right|_{z=-0} + Ew_0^n + M = \left.\frac{\partial w_2^n}{\partial z}\right|_{z=+0}, \quad w_0^n = w_1^n|_{z=0},$$
$$\xi w_1^n|_{z=-0} = w_2^n|_{z=+0}, \tag{4.24}$$

$$\left[\frac{\partial w_1^{n+1}}{\partial z} + s_1(t,-a_1)w_1^{n+1}\right]_{z=-a_1} = q_1(t,-a_1), \tag{4.25}$$

$$\left[\frac{\partial w_2^{n+1}}{\partial z} + s_2(t,a_2)w_2^{n+1}\right]_{z=a_2} = q_2(t,a_2). \tag{4.26}$$

Let's divide the received balance problem (4.22)–(4.26) into two different problems and each solution will be found by means of the Laplace integral transformation with infinite limits on a spatial variable. It is supposed that in calculation of images on coordinate $z$ we operate with the functions analytically continued on values $z < -a_1$, $z > a_2$ by that law which they are defined in an interval $-a_1 < z < a_2$.

Let's apply the Laplace integral transformation [6] to the differential equations from Eq. (4.22) separately, excluding derivative on $z$ and replacing it with linear expression in relation to the required function. Then we consider the functions for which the Laplace integral transformation converges absolutely. The valid part of complex number $p$ is considered positive, that is $\mathrm{Re}\, p > 0$. We will introduce capital letters for images $W_1, H_1$ and omit, for simplicity of calculations, an index $n$ above at $w_1$. We will have:

$$W_1(t,p) = \int_{-\infty}^{0} \exp(pz) w_1(t,z)\, dz, \quad w_1(t,z) = L^{-1}[W_1(t,p)],$$
$$h_1(t,z) = L^{-1}[H_1(t,p)], \quad -\infty < z \le 0. \tag{4.27}$$

In expression (4.27) by replacement of a variable $z = -x$ we transform the region $-\infty < z \le 0$ into the region $0 \le x < \infty$ then using formulas from [6], we will have $\partial g/\partial x = \partial w_1(t,0)/\partial x$, $g = w_1(t,0)$:

$$\frac{\partial w_1}{\partial x} = L^{-1}[pW_1(t,p) - g],$$
$$\frac{\partial^2 w_1}{\partial x^2} = L^{-1}\left[p^2 W_1(t,p) - pg - \frac{\partial g}{\partial x}\right], \quad 0 \le x < \infty.$$

We assume that the desired solution $w_1(t,x)$, and also its derivatives entering into the Eq. (4.22), satisfy the Laplace integral transformation conditions on $x$, and besides, growth degree on $x$ functions $w_1(t,x)$ and its derivatives does not depend on $t$. Multiplying both members of Eq. (4.22) on $\exp(-px)$ and integrating on $x$ from 0 to $\infty$ we will receive [6, 16], $\Phi_1$ obviously independent on $x$ in Eq. (4.20) [$\Phi_1$ is always possible to set on the bottom iteration on $n$ knowing values $w_1^n$ in initial and the subsequent moments of time from formulas (4.14) and (4.17)]:

$$p^2 W_1(t,p) - pg - \frac{\partial g}{\partial x} + \Phi_1 W_1(t,p) = H_1(t,p) \quad \text{or}$$

$$W_1 = \frac{pg}{p^2 + b_1^2} + \frac{b_1 \frac{\partial g}{\partial x}}{b_1(p^2 + b_1^2)} + \frac{b_1 H_1}{b_1(p^2 + b_1^2)}, \quad b_1 = \sqrt{\Phi_1}. \tag{4.28}$$

Using the Laplace integral transformation [6]: $L^{-1}[p/(p^2 + b_1^2)] = \cos(b_1 x)$ at $b_1^2 > 0$, $L^{-1}[p/(p^2 - b_1^2)] = \cosh(b_1 x)$ at $b_1^2 < 0$; $L^{-1}[H_1(p)/p] = \int_0^x h_1(y)\,dy$, we restore the original for $w_1(t,x)$ from Eq. (4.28) [6]

$$w_1(t,x) = g u_1(x) + u_2(x)\frac{\partial g}{\partial x} + \int_0^x u_2(x-y)h_1(y)\,dy, \tag{4.29}$$

where $u_1(x) = \cos(b_1 x), u_2(x-y) = b_1^{-1}\sin[b_1(x-y)], u_2(x) = b_1^{-1} \times \sin(b_1 x)$, at $b_1^2 = \Phi_1 > 0$ from Eq. (4.20); $u_1(x) = \cosh(b_1 x), u_2(x) = b_1^{-1}\sinh(b_1 x), u_2(x-y) = b_1^{-1}\sinh[b_1(x-y)]$ at $b_1^2 < 0$.

Similarly for $i = 2$ according to the algorithm (4.22)–(4.29) in the range of $0 < z < \infty$ we receive at $b_2 = \sqrt{\Phi_2}$ if using the formula (4.24) and notation (4.21) for $h_2$

$$w_2(t,z) = g[\xi u_1(z) + u_2(z)E] + u_2(z)\left[\eta\frac{\partial g}{\partial x} + M\right] + \int_0^z u_2(z-y)h_2(y)\,dy, \tag{4.30}$$

where $u_1(z) = \cos(b_2 z), u_2(z-y) = b_2^{-1}\sin[b_2(z-y)], u_2(z) = b_2^{-1} \times \sin(b_2 z)$, at $b_2^2 = \Phi_2 > 0$ from Eq. (4.20) $u_2(z) = \cosh(b_2 z), u_2(z) = b_2^{-1}\sinh(b_2 z), u_2(z-y) = b_2^{-1}\sinh[b_2(z-y)]$ at $b_2^2 < 0$.

To find the unknown derivative $\partial g/\partial x$ and functions $g$ in Eqs. (4.29) and (4.30) we differentiate the last one on $x$ and $z$ [it is supposed that there is a limited partial derivative on $x$, $z$ from $w_1(t,x)$, $w_2(t,z)$]. Then we will receive at $b_i = \sqrt{\Phi_i}, i = 1,2$ from Eq. (4.20), obviously not dependent on $x$, $z$

$$\begin{aligned}\frac{\partial w_1}{\partial x} &= \xi_1 g u_2(x) + u_1(x)\frac{\partial g}{\partial x} + \int_0^x u_1(x-y)h_1(y)\,dy,\\ \frac{\partial w_2}{\partial z} &= g[\xi\xi_2 u_2(z) + u_1(z)E] + u_1(z)\left[\eta\frac{\partial g}{\partial x} + M\right]\\ &\quad + \int_0^z u_1(z-y)h_2(y)\,dy, \quad \xi_i = \operatorname{sign} b_i^2 = \mp b_i^2, \quad i = 1,2,\end{aligned} \tag{4.31}$$

where the sign "minus" is taken at $\Phi_i > 0, i = 1,2$, and "plus"—at $\Phi_i < 0$, $i = 1,2$ from Eq. (4.20).

Finally we find a derivative $\partial g/\partial x$ and function $g$ if we use the Eqs. (4.24)–(4.26) and Eqs. (4.29)–(4.31):

$$\left.\frac{\partial w_1}{\partial x}\right|_{x=a_1} = q_1(t,a_1) - s_1(t,a_1)w|_{x=a_1} = g\xi_1(a_1)u_2(a_1)$$
$$+ u_1(a_1)\frac{\partial g}{\partial x} + \int_0^{a_1} u_1(a_1-y)h_1(y)\,dy, \tag{4.32}$$

$$\left.\frac{\partial w_2}{\partial z}\right|_{z=a_2} = q_2(t,a_2) - s_2(t,a_2)w|_{z=a_2} = g[\xi\xi_2(a_2)u_2(a_2)$$
$$+ u_2(a_2)E] + u_1(a_2)\left(\eta\frac{\partial g}{\partial x} + M\right) + \int_0^{a_2} u_1(a_2-y)h_2(y)\,dy, \tag{4.33}$$

$$w_1(a_1) = gu_1(a_1) + u_2(a_1)\frac{\partial g}{\partial x} + \int_0^{a_1} u_2(a_1-y)h_1(y)\,dy. \tag{4.34}$$

$$w_2(a_2) = g[\xi u_1(a_2) + u_2(a_2)E] + u_2(a_2)\left(\eta\frac{\partial g}{\partial x} + M\right)$$
$$+ \int_0^{a_2} u_2(a_2-y)h_2(y)\,dy. \tag{4.35}$$

Substituting $w_1(a_1)$ from Eq. (4.34) and $w_2(a_2)$ from Eq. (4.35) in the Eqs. (4.32) and (4.33) we find

$$q_1(t,a_1) - s_1(t,a_1)\left[gu_1(a_1) + u_2(a_1)\frac{\partial g}{\partial x} + \int_0^{a_1} u_2(a_1-y)h_1(y)\,dy\right]$$
$$= g\xi_1(a_1)u_2(a_1) + u_1(a_1)\frac{\partial g}{\partial x} + \int_0^{a_1} u_1(a_1-y)h_1(y)\,dy, \tag{4.36}$$

$$q_2(t,a_2) - s_2(t,a_2)\left\{g[\xi u_1(a_2) + u_2(a_2)E] + u_2(a_2)\left(\eta\frac{\partial g}{\partial x} + M\right)\right.$$
$$\left.+ \int_0^{a_2} u_2(a_2-y)h_2(y)\,dy\right\} = g[\xi\xi_2(a_2)u_2(a_2) + u_1(a_2)E]$$
$$+ u_1(a_2)\left(\eta\frac{\partial g}{\partial x} + M\right) + \int_0^{a_2} u_1(a_2-y)h_2(y)\,dy. \tag{4.37}$$

Finally we receive from Eqs. (4.36) and (4.37)

$$b_1g + b_2\frac{\partial g}{\partial x} = b_3, \quad c_1g + c_2\frac{\partial g}{\partial x} = c_3, \tag{4.38}$$

$$b_1 = s_1(a_1)u_1(a_1) + \xi_1(a_1)u_2(a_1), b_2 = u_1(a_1) + s_1(a_1)u_2(a_1),$$
$$b_3 = q_1(a_1) - \int_0^{a_1} [u_1(a_1-y) + s_1(a_1)u_2(a_1-y)]h_1(y)\,dy,$$

$$
\begin{aligned}
c_1 &= Eu_1(a_2) + \xi\xi_2(a_2)u_2(a_2) + s_2(a_2)[Eu_2(a_2) + \xi u_1(a_2)],\\
c_2 &= \eta[u_1(a_2) + s_2(a_2)u_2(a_2)],\\
c_3 &= q_2(a_2) - \int_0^{a_2} [u_1(a_2 - y) + s_2(a_2)u_2(a_2 - y)]h_2(y)\,dy\\
&\quad - M[u_1(a_2) + s_2(a_2)u_2(a_2)].
\end{aligned}
\tag{4.39}
$$

Further solving the system of the equations from Eq. (4.38), we have

$$
\begin{aligned}
g &= \frac{b_3c_2 - b_2c_3}{\Delta}, \quad \frac{\partial g}{\partial x} = \frac{b_1c_3 - b_3c_1}{\Delta},\\
\Delta &= b_1c_2 - c_1b_2 = [s_1u_1(a_1) + \xi_1(a_1)u_2(a_1)]\eta[u_1(a_2)\\
&\quad + s_2(a_2)u_2(a_2)] - \{Eu_1(a_2) + \xi\xi_2(a_2)u_2(a_2)\\
&\quad + s_2(a_2)[Eu_2(a_2) + \xi u_1(a_2)]\}[u_1(a_1) + s_1(a_1)u_2(a_1)].
\end{aligned}
\tag{4.40}
$$

Substituting $g$ and $\partial g/\partial x$ from Eq. (4.40) in the Eqs. (4.29) and (4.30) at $x = 0$, designating $w_1(t, 0) = w_0$ and taking into consideration Eqs. (4.39) and (4.40) we have:

$$
\dot{w}_0^{n+1} + U_0 w_0^{n+1} = X_0, \quad n = 0, 1, 2, \ldots, \tag{4.41}
$$

$$
\begin{aligned}
X_0 &= (K_0 + Y_0R_0)/(C_0Y_0), \quad C_0 = 0.5(C_1 + C_2)_0/A_0,\\
R_0 &= (R_1 + R_2)_0/2,\\
U_0 &= 1/(Y_0C_0), \quad K_0 = \Delta^{-1}q_1\eta[u_1(a_2) + s_2(a_2)u_2(a_2)]\\
&\quad + \Delta^{-1}[u_1(a_1) + s_1(a_1)u_2(a_1)]\{M[u_1(a_2) + s_2(a_2)u_2(a_2)] - q_2\},\\
Y_0 &= \frac{1}{\Delta}\left\{\eta[u_1(a_2) + s_2(a_2)u_2(a_2)]\int_0^{a_1}[s_1(a_1)u_2(a_1 - y) + u_1(a_1 - y)]\,dy\right.\\
&\quad \left. -[u_1(a_1) + s_1(a_1)u_2(a_1)]\int_0^{a_2}[s_2(a_2)u_2(a_2 - y) + u_1(a_2 - y)]\,dy\right\};
\end{aligned}
\tag{4.42}
$$

$$
\begin{aligned}
w_1^{n+1} &= -\int_0^{a_1} G_1(x, y)h_1(y)\,dy + V_1(h_2) + K_1(x),\\
h_1 &= c_1\dot{w}_1^{n+1} - R_1(w_1^n),
\end{aligned}
\tag{4.43}
$$

$$
G_1(x, y) = \begin{cases} P_1(x)[s_1(a_1)u_2(a_2 - y) + u_1(a_1 - y)] - u_2(x - y), & 0 \le y \le x,\\ P_1(x)[s_1(a_1)u_2(a_2 - y) + u_1(a_1 - y)], & x \le y \le a_1, \end{cases}
\tag{4.44}
$$

$$\begin{aligned}
&\dot{w}_1^{n+1} + U_1 w_1^{n+1} = X_1,\\
&X_1 = Y_1^{-1}\left[\int_0^{a_1} G_1(x,y)R_1(w_1^n)\,dy + K_1(x) + V_1(h_2)\right],\\
&n = 0,1,2,\ldots,
\end{aligned} \tag{4.45}$$

$$\begin{aligned}
Y_1 &= C_1\int_0^{a_1}\frac{G_1(x,y)}{A_1}\,dy,\quad U_1 = Y_1^{-1},\quad K_1(x) = \Delta^{-1}u_1(x)\\
&\times\{q_1\eta[u_1(a_2)+s_2(a_2)u_2(a_2)] - [u_1(a_1)+s_1(a_1)u_2(a_1)][q_2\\
&- M(u_1(a_2)+s_2(a_2)u_2(a_2))]\} + \Delta^{-1}u_2(x)\{[s_1(a_1)u_1(a_1)\\
&+ \xi_1(a_1)u_2(a_1)][q_2 - M(u_1(a_2)+s_2(a_2)u_2(a_2))] - q_1[Eu_1(a_2)\\
&+ \xi\xi_2(a_2)u_2(a_2) + s_2(a_2)(Eu_2(a_2)+\xi u_1(a_2))]\},\\
V_1(h_2) &= \int_0^{a_2}[s_2(a_2)u_2(a_2-y)+u_1(a_2-y)]h_2\,dy\{u_1(x)[u_1(a_1)\\
&+ s_1(a_1)u_2(a_1)] - u_2(x)[s_1(a_1)u_1(a_1)+\xi_1(a_1)u_2(a_1)]\}/\Delta,\\
P_1(x) &= \Delta^{-1}u_1(x)\eta[(u_1(a_2)+s_2(a_2)u_2(a_2)] - \Delta^{-1}u_2(x)\\
&\times\{Eu_1(a_2)+\xi\xi_2(a_2)u_2(a_2)+s_2(a_2)[Eu_2(a_2)+\xi u_1(a_2)]\},\\
h_2 &= [V_2(h_1)+K_2(z)-w_2^{n+1}]/\int_0^{a_2}G_2(z,y)\,dy;
\end{aligned} \tag{4.46}$$

$$\begin{aligned}
w_2^{n+1} &= -\int_0^{a_2}G_2(z,y)h_2(y)\,dy + V_2(h_1) + K_2(z),\\
h_2 &= c_2\dot{w}_2^{n+1} - R_2(w_2^n),
\end{aligned} \tag{4.47}$$

$$G_2(z,y)=\begin{cases}P_2(z)[s_2(a_2)u_2(a_2-y)+u_1(a_2-y)] - u_2(z-y), & 0\le y\le z,\\ P_2(z)[s_2(a_2)u_2(a_2-y)+u_1(a_2-y)], & z\le y\le a_2,\end{cases} \tag{4.48}$$

$$\begin{aligned}
&\dot{w}_2^{n+1} + U_2 w_2^{n+1} = X_2,\\
&X_2 = Y_2^{-1}\left[\int_0^{a_2} G_2(x,y)R_2(w_2^n)\,dy + K_2(z) \quad + V_2(h_1)\right],\\
&n = 0,1,2,\ldots,
\end{aligned} \tag{4.49}$$

$$\begin{aligned}
Y_2 &= C_2\int_0^{a_2}\frac{G_2(z,y)}{A_2}\,dy,\quad U_2 = Y_2^{-1},\quad K_2(z) = [u_1(z)+u_2(z)E]\\
&\times\{q_1\eta[u_1(a_2)+s_2(a_2)u_2(a_2)] - [u_1(a_1)+s_1(a_1)u_2(a_1)][q_2\\
&- M(u_1(a_2)+s_2(a_2)u_2(a_2))]\}/\Delta + \eta u_2(z)\{[s_1(a_1)u_1(a_1)
\end{aligned}$$

$$+ \xi_1(a_1)u_2(a_1)][q_2 - M(u_1(a_2) + s_2(a_2)u_2(a_2))] - q_1[Eu_1(a_2) + \xi\xi_2(a_2)u_2(a_2) + s_2(a_2)(Eu_2(a_2) + \xi u_1(a_2))]\}/\Delta + u_2(z)M,$$

$$V_2(h_1) = \int_0^{a_1} [s_1(a_1)u_2(a_1 - y) + u_1(a_1 - y)]h_1\, dy\{\eta u_2(z) \times [\xi\xi_2(a_2)u_2(a_2) + u_1(a_2)E + s_2(a_2)(\xi u_1(a_2) + u_2(a_2)E)] - \eta[\xi u_1(z) + u_2(z)E][s_2(a_2)u_2(a_2) + u_1(a_2)]\}/\Delta,$$

$$P_2(z) = \Delta^{-1}\{u_2(z)\eta[\xi_1(a_1)u_2(a_1) + s_1(a_1)u_1(a_1)] - [Eu_2(z) + \xi u_1(z)][u_1(a_1) + s_1(a_1)u_2(a_1)]\},$$

$$h_1 = [V_1(h_2) + K_1(x) - w_1^{n+1}]/\int_0^{a_1} G_1(x, y)\, dy. \tag{4.50}$$

As a result the final solution of problems (4.41)–(4.50) will be [33]:

$$w_0^{n+1}(0, t) = \left[w_{H,0} + \int_0^t X_0(w_0^n, \tau)\exp(\tau U_0)\, d\tau\right] \times \exp(-tU_0), \quad n = 0, 1, 2, \ldots, \tag{4.51}$$

$$w_1^{n+1}(x, t) = \left[w_{H,1} + \int_0^t X_1(w_1^n, x, \tau)\exp(\tau U_1)\, d\tau\right] \times \exp(-tU_1), \quad 0 < x \le a_1, \quad n = 0, 1, 2, \ldots, \tag{4.52}$$

$$w_2^{n+1}(z, t) = \left[w_{H,2} + \int_0^t X_2(w_2^n, z, \tau)\exp(\tau U_2)\, d\tau\right] \times \exp(-tU_2), \quad 0 < z \le a_2, \quad n = 0, 1, 2, \ldots, \tag{4.53}$$

the solution $T_i(z, t), i = 1, 2$ of the initial conjugate boundary problem (4.1)–(4.6) is possible due to inversion formulas (4.17).

For simplicity of further analysis and comparison of accuracy of the received formulas with the known analytical solution [45] we will write working solutions (4.51)–(4.53) for boundary conditions of the first type in the absence of sources: $B_i = Y_i = E_i = 0$ and $C_i = \text{const}, A_i = \text{const}, i = 1, 2$, then the Eqs. (4.42), (4.46), and (4.50) will be

$$K_0 = \Delta^{-1}\{q_1(a_1)a_2\eta - a_1[q_2(a_2) - a_2M]\},$$

$$\Delta = a_2\eta - a_1, \quad \eta = A_1/A_2,$$

$$Y_0 = \Delta^{-1}\left[a_2\eta \int_0^{a_1}(a_1 - y)\, dy - a_1 \int_0^{a_2}(a_2 - y)\, dy\right], \quad U_0 = (YC)_0^{-1},$$

$$C_0 = (C_1 + C_2)_0/2, \quad X_0 = K_0/(Y_0C_0), \tag{4.54}$$

$$K_1(x) = \Delta^{-1}\{q_1(a_1)(a_2\eta - x) + [q_2(a_2) - a_2M](x - a_1)\},$$

$$V_1(h_2) = \frac{a_1 - x}{\Delta}\int_0^{a_2}(a_2 - y)h_2(y)\,dy, h_2 = [V_2(h_1) + K_2(z)$$
$$- w_2^{n+1}]/\int_0^{a_2} G_2(z, y)dy, \quad Y_1 = \frac{C_1}{A_1}\int_0^{a_1} G_1(x, y)\,dy,$$

$$G_1(x, y) = \begin{cases}(a_2\eta - x)(a_1 - y)/\Delta - (x - y), & 0 \le y \le x,\\ (a_2\eta - x)(a_1 - y)/\Delta, & x \le y \le a_1,\end{cases}$$

$$X_1 = Y_1^{-1}[K_1(x) + V_1(h_2)], \quad U_1 = Y_1^{-1}, \tag{4.55}$$

$$K_2(z) = \Delta^{-1}\{q_1(a_1)(a_2 - z)\eta + [q_2(a_2) - a_2M](z\eta - a_1)\} + zM,$$
$$V_2(h_1) = \frac{\eta(z - a_2)}{\Delta}\int_0^{a_1}(a_1 - y)h_1(y)\,dy, h_1 = [V_1(h_2) + K_1(x)$$
$$- w_1^{n+1}]/\int_0^{a_1} G_1(x, y)\,dy, \quad Y_2 = \frac{C_2}{A_2}\int_0^{a_2} G_2(z, y)\,dy,$$

$$G_2(z, y) = \begin{cases}(z\eta - a_1)(a_2 - y)/\Delta - (z - y), & 0 \le y \le z,\\ (z\eta - a_1)(a_2 - y)/\Delta, & z \le y \le a_2,\end{cases}$$

$$X_2 = Y_2^{-1}[K_2(z) + V_2(h_1)], \quad U_2 = Y_2^{-1}. \tag{4.56}$$

As a result, types of solutions (4.41), (4.45), (4.49), (4.51)–(4.53) with the inversion formula (4.12) and absence of iterations will stay the same. That is universalism of the developed mathematical technology.

In [45] the analytical solution of ignition of a wooden wall as a result of fire is given. The equation of heat conductivity for a zone of a fire and a layer of wood looks like [45]

$$\frac{\partial T_1}{\partial t} = \xi_1\frac{\partial^2 T_1}{\partial x^2}, \quad -\infty < x < 0, \qquad \frac{\partial T_2}{\partial t} = \xi_2\frac{\partial^2 T_2}{\partial x^2}, \quad 0 < x < \infty, \tag{4.57}$$

where $\xi_i, i = 1, 2$ are coefficients of thermal diffusivity of natural fire and a wooden wall, $T_1$, $T_2$ are temperatures in a zone of fire and in a wood layer.

In a wood-fire interface, the conditions of equality of temperatures of fire and wood are satisfied and the balance of thermal energy taking into account oxidation and evaporation of water from it is:

$$\lambda_1 \frac{\partial T_1}{\partial x}\bigg|_{x=-0} + \epsilon\sigma T_\Gamma^4 - \frac{q_1 A(P_w - P_e)}{\sqrt{2\pi R T_w/M}} + q_2 k_2 \rho_w c_w \exp\left(-\frac{E_2}{RT_w}\right)$$
$$= \lambda_2 \frac{\partial T_2}{\partial x}\bigg|_{x=+0}, \quad T_1|_{x=-0} = T_2|_{x=+0}, \tag{4.58}$$

where $P_w = k_1 \exp(-E_1/RT_w)$, $T_\Gamma$ is a temperature of burning in front fire, $\epsilon$ a coefficient of blackness, $\sigma$ is a Stefan-Boltzmann constant, $R$ is a universal gas constant, $M$ is molecular weight, $E_i, q_i, k_i, i = 1, 2$ are energy activation, preexhibitors, and thermal effects of reactions of evaporation and oxidation, $\rho_w$ is density of a gas phase on a wood surface, $c_w$ is concentration of oxygen on a wood surface, $\lambda_1, \lambda_2$ are coefficients of heat conductivity of the air environment and wood, $P_e$ is pressure in an environment, $A$ is a constant, $T_w$ is temperature of surface of wood, $T_H$ is initial temperature.

While taking into account initial and boundary conditions:

$$T_1|_{t=0} = T_\Gamma, \quad T_2|_{t=0} = T_H, \quad \lim_{x\to-\infty} T_1 = T_\Gamma, \quad \lim_{x\to\infty} T_2 = T_H \tag{4.59}$$

in [45] the analytical solution (automodeling) of problems (4.57)–(4.59) is found

$$T_1 = T_\Gamma + B(t, T_\Gamma, T_H)\left[ERF\left(\frac{|x|}{2\sqrt{\xi_1 t}}\right) - 1\right], \tag{4.60}$$

$$T_2 = T_\Gamma - B(t, T_\Gamma, T_H) + ERF\left(\frac{x}{2\sqrt{\xi_2 t}}\right)[B(t, T_\Gamma, T_H) - T_\Gamma + T_H], \tag{4.61}$$

$$B(t, T_\Gamma, T_H) = \frac{\sqrt{\pi\xi_1\xi_2 t}}{\lambda_1\sqrt{\xi_2} + \lambda_2\sqrt{\xi_1}}\left[\epsilon\sigma T_\Gamma^4 + \frac{\lambda_2(T_\Gamma - T_H)}{\sqrt{\pi\xi_2 t}}\right.$$
$$\left. - \frac{q_1 A(P_w - P_e)}{\sqrt{2\pi R T_w/M}} + q_2 k_2 \rho_w c_w \exp\left(-\frac{E_2}{RT_w}\right)\right],$$
$$c_w = \frac{\sqrt{D} c_H}{\sqrt{D} + k_2\sqrt{\pi t}\ \exp(-E_2/RT_w)}, \tag{4.62}$$

where $D$ is a coefficient of diffusion, $ERF\left(\frac{x}{2\sqrt{\xi t}}\right)$ is a function of errors [30, 42].

## Existence, Uniqueness, and Convergence

Without loss of generality, we will consider the solution of the simplified boundary problem (4.1)–(4.4) with zero boundary conditions of the first type.

We take a nonlinear case $m > 0$, $k > 1$ in the region $\overline{Q_i} = Q_i + \Gamma$, $\overline{Q_{t,i}} = \overline{Q_i} \times [0 < t \leq t_0]$ at $Y_i = 1, C_i = 1, A_i(T_i) = T_i^m, A_{H,i} = 1$, $N_i = 1, E_i = 0, i = 1,2$

$$\frac{\partial T_i}{\partial t} = \frac{\partial}{\partial z}\left[A_i(T_i)\frac{\partial T_i}{\partial z}\right] + B_i(T_i)\frac{\partial T_i}{\partial z} + T_i^k,$$
$$i = 1 : -a_1 < z < 0, \quad i = 2 : 0 < z < a_2, \quad 0 < t < t_0,$$
$$T_i(0,z) = T_{H,i}(z), \quad T_i|_\Gamma = 0, \quad i = 1,2,$$
$$A_1 \frac{\partial T_1}{\partial z}\bigg|_{z=-0} = A_2 \frac{\partial T_2}{\partial z}\bigg|_{z=+0}, \quad T_1|_{z=-0} = T_2|_{z=+0}. \tag{4.63}$$

After application of the Kirhgof's transformation (4.5) at $A_{H,i} = 1, i = 1,2$ and formulas (4.7) and (4.17) we have the modified boundary problem

$$c_i(w_i)\frac{\partial w_i}{\partial t} = \frac{\partial^2 w_i}{\partial z^2} + \exp(r_i z)[s w_i \exp(-r_i z)]^{k/s}$$
$$- w_i r_i^2, \quad i = 1,2, \quad 0 < t < t_0,$$
$$w_i|_{t=0} = w_{H,i}, \quad w_{H,i} = \frac{T_{H,i}^s \exp(r_i z)}{s}, \quad w_i|_\Gamma = 0,$$
$$r_i = 0.5 B_i / A_i, \quad i = 1,2,$$
$$\eta \frac{\partial w_1}{\partial z}\bigg|_{z=-0} + E w_0 = \frac{\partial w_2}{\partial z}\bigg|_{z=+0}, \quad \xi w_1|_{z=-0} = w_2|_{z=+0}. \tag{4.64}$$

Thus $T_i, i = 0,1,2$ a boundary problem (4.63) is defined from Eq. (4.17) according to the inversion formula:

$$T_i = [w_i s \exp(-r_i z)]^{1/s}, \quad i = 1,2, \quad T_0 = (s w_0)^{1/s}. \tag{4.65}$$

As a result of the algorithm application (4.18)–(4.24), (4.28)–(4.53) to a boundary problem (4.64) we have

$$w_i^{n+1}(z,t) = \left\{ w_{H,i} + \int_0^t F_i \exp[\tau U_i(w_i^n)]\, d\tau \right\}$$
$$\times \exp[-t U_i(w_i^n)], \quad i = 1,2, \quad -a_1 < z < a_2,\ z \neq 0, \tag{4.66}$$

$$w_0^{n+1}(0,t) = \left\{ w_{H,0} + \int_0^t F_0 \exp[\tau U_0(w_0^n)]\, d\tau \right\} \\ \times \exp[-tU_0(w_0^n)], \quad n = 0, 1, 2, \ldots, \tag{4.67}$$

$$F_i = Y_i^{-1}\left[ V_i + \int_0^{a_i} G_i(z,y) R_i(w_i^n)\, dy \right], \quad \Phi_i = kZ_i^{k/s-1} - r_i^2,$$
$$R_i = Z_i^{k/s}[\exp(r_i z) - w_i^n k/Z_i], \quad Z_i = sw_i^n \exp(-r_i z),$$
$$Y_i = \int_0^{a_i} c_i(w_i^n) G_i(z,y)\, dy, \quad c_i = A_i^{-1}(w_i^n), \quad U_i = Y_i^{-1}, \quad i = 1, 2,$$

$$G_i(z,y) = \begin{cases} P_i(z,y) - u_2(z-y), & i = 1,2, \quad 0 \le y \le z, \\ P_i(z,y), \quad z \le y \le a_i, & i = 1,2, \end{cases} \tag{4.68}$$

$$P_1(x,y) = \{\eta u_1(x) u_2(a_2) - u_2(x)[\xi u_1(a_2) + Eu_2(a_2)]\} \\ \times u_2(a_1 - y)/\Delta,$$
$$P_2(z,y) = \{\eta u_2(z) u_1(a_1) - [\xi u_1(z) + Eu_2(z)] u_2(a_1)\} \\ \times u_2(a_2 - y)/\Delta,$$
$$\Delta = u_1(a_1)\eta u_2(a_2) - u_2(a_1)[\xi u_1(a_2) + Eu_2(a_2)], \tag{4.69}$$

$$V_1^{(1)}(h_2^{(1)}) = \Delta^{-1}[u_1(z)u_2(a_1) - u_2(z)u_1(a_1)] \int_0^{a_2} u_2(a_2 - y) h_2^{(1)}\, dy,$$

$$V_2^{(1)}(h_1^{(1)}) = \Delta^{-1}\eta\{u_2(x_1)[\xi u_1(a_2) + u_2(a_2)E] \\ - u_2(a_2)[\xi u_1(x_1) + u_2(x_1)E]\} \int_0^{a_1} u_2(a_1 - y) h_1^{(1)}\, dy,$$

$$F_0 = (Y_0 c_0)^{-1}(K_0 + Y_0 R_0), \quad K_0 = 0, \quad R_0 = (Z_0^n)^{k/s}\left(1 - \frac{w_0^n k}{Z_0^n}\right),$$
$$U_0 = (Y_0 c_0)^{-1}, \quad c_0 = A_0^{-1}(w_0^n), \quad Z_0^n = sw_0^n,$$
$$Y_0 = \left[ \eta u_2(a_2) \int_0^{a_1} u_2(a_1 - y)\, dy - u_2(a_1) \int_0^{a_2} u_2(a_2 - y)\, dy \right] / \Delta. \tag{4.70}$$

The final solution $T_i, i = 0, 1, 2$ of a boundary problem (4.63) is received after substitutions $w_i^{n+1}, i = 0, 1, 2$ defining by formulas (4.66)–(4.70) in inversion formulas (4.65).

**Theorem.** *Let $w_i, i = 0, 1, 2$ be continuously differentiated in $\overline{Q_{t,i}}, i = 1, 2$, then in regions $\overline{Q_{t,i}}, i = 1, 2$ there is a unique solution of the modified boundary problem* (4.64).

Existence and uniqueness of the solution of a boundary problem (4.63) are proved in the same manner as in Chapter 2 [see formulas (2.19)–(2.30)].

## Estimation of Speed of Convergence [15, 16]

It is considered that in some neighborhood of a root function $f_i = f_i(w_i^n, \dot{w}_i^n), i = 1, 2$ from Eq. (4.18) together with the partial derivatives $\partial f_i/\partial w_i$, $\partial^2 f_i/\partial w_i^2, \partial f_i/\partial \dot{w}_i, \partial^2 f_i/\partial \dot{w}_i^2, i = 1, 2$ are continuous, and $\partial f_i/\partial w_i$, $\partial^2 f_i/\partial w_i^2$, $\partial f_i/\partial \dot{w}_i$, $\partial^2 f_i/\partial \dot{w}_i^2, i = 1, 2$ in this neighborhood does not go to zero.

Let's address the recurrence relationship (4.18) and noticing that $f_i(w_i, \dot{w}_i) = s_i(w_i) + r_i(\dot{w}_i), i = 1, 2$ in Eq. (4.18) we will subtract $n$-e the equation from $(n+1)$th then we will find:

$$\frac{\partial^2(w_i^{n+1} - w_i^n)}{\partial z^2} = s_i(w_i^n) - s_i(w_i^{n-1}) - (w_i^n - w_i^{n-1})\frac{\partial s_i(w_i^{n-1})}{\partial w_i^{n-1}}$$
$$+ (w_i^{n+1} - w_i^n)\frac{\partial s_i(w_i^n)}{\partial w_i^n} + \left[ r_i(\dot{w}_i^n) - r_i(\dot{w}_i^{n-1}) \right.$$
$$\left. -(\dot{w}_i^n - \dot{w}_i^{n-1})\frac{\partial r_i(\dot{w}_i^{n-1})}{\partial \dot{w}_i^{n-1}} + (\dot{w}_i^{n+1} - \dot{w}_i^n)\frac{\partial r_i(\dot{w}_i^n)}{\partial \dot{w}_i^n} \right], \quad i = 1, 2. \quad (4.71)$$

From the average theorem [33] it follows:

$$s_i(w_i^n) - s_i(w_i^{n-1}) - (w_i^n - w_i^{n-1})\frac{\partial s_i(w_i^{n-1})}{\partial w_i^{n-1}} = 0.5(w_i^n - w_i^{n-1})^2$$
$$\times \frac{\partial^2 s_i(\xi_i)}{\partial w_i^2}, \quad w_i^{n-1} \le \xi_i \le w_i^n, i = 1, 2.$$

Let's consider Eq. (4.71) how the equation is in relation to $u_i^{n+1} = w_i^{n+1} - w_i^n$, $i = 1, 2$ and transform it as above Eqs. (4.28)–(4.48). Then we will have:

$$\frac{\partial^2 u_i^{n+1}}{\partial z^2} + u_i^{n+1}\Phi_i = h_i, \quad h_i = \dot{u}_i^{n+1}\frac{\partial r_i(\dot{w}_i^n)}{\partial \dot{w}_i^n} + F_i, \quad \Phi_i = -\frac{\partial s_i(w_i^n)}{\partial w_i^n},$$
$$F_i = 0.5\left[ (u_i^n)^2\frac{\partial^2 s_i(w_i^n)}{\partial w_i^2} + (\dot{u}_i^n)^2\frac{\partial^2 r_i(\dot{w}_i^n)}{\partial \dot{w}_i^2} \right], \quad u_i(0, x) = 0, \quad i = 1, 2,$$

$$\eta \frac{\partial u_1}{\partial z}\bigg|_{z=-0} + Eu_0 = \frac{\partial u_2}{\partial z}\bigg|_{z=+0}, \quad \xi u_1|_{z=-0} = u_2|_{z=+0}, \quad u_i|_\Gamma = 0,$$

$$u_i^{n+1} = \int_0^{a_i} G_i(z,y)h_i\,dy, \quad K_i(z) = 0, \quad \frac{\partial^2 r_i(\dot{w}_i^n)}{\partial \dot{w}_i^2} = 0, \quad i = 1,2,$$

$$\dot{u}_i^{n+1} - u_i^{n+1}\alpha_i = -F_i/Y_i, \quad \alpha_i = 1/Y_i, \quad V_i(a_i) = 0,$$

$$Y_i = \int_0^{a_i} G_i(z,y)\frac{\partial r_i(\dot{w}_i^n)}{\partial \dot{w}_i^n}\,dy, \quad i = 1,2, \tag{4.72}$$

where $G_i(z,y)$ are taken from Eq. (4.68) at $c_i = \frac{\partial r_i(\dot{w}_i^n)}{\partial \dot{w}_i^n}, i = 1,2$ for $Y_i$.

Finally the problem solution (4.72) will be Eq. (4.66), where $w_{H,i} = 0$, $i = 1,2$.

Let's put $\max\limits_{w_i,\dot{w}_i \in \overline{Q_{t,i}}}\left(\left|\frac{\partial^2 s_i(w_i^n)}{\partial w_i^2}\right|, \left|\frac{\partial^2 r_i(\dot{w}_i^n)}{\partial \dot{w}_i^2}\right|\right) = c_2$, $\max\limits_{w_i \in \overline{Q_{t,i}}} |\Phi_i| = c_3$, $\max\limits_{w_i,\dot{w}_i \in \overline{Q_{t,i}}}\left(\left|\frac{\partial s_i(w_i^n)}{\partial w_i^n}\right|, \left|\frac{\partial r_i(\dot{w}_i{}^n)}{\partial \dot{w}_i{}^n}\right|\right) = c_1$, $\max\limits_{z,y}|G(z,y)| \le |b_i^{-1}\sin(b_i y)| \le b_i^{-1}$, $b_i^2 = \Phi_i, i = 1,2$, assuming that $c_m < \infty, m = 1,2,3$. Then from the Eqs. (4.66) and (4.72) it follows, if we introduce $B = c_2/2c_1$, $Y_i = c_1 a_i/\sqrt{c_3}$, $\alpha_i = \sqrt{c_3}/(c_1 a_i), i = 1,2$:

$$|u_i^{n+1}| \le B\exp(t\alpha_i)\int_0^t (u_i^n)^2 \exp(-\alpha_i \tau)\,d\tau, \quad i = 1,2. \tag{4.73}$$

Let's choose $u_i^0(t,z)$ so that $|u_i^0(t,z)| \le 1$ is in the region $\overline{Q_{t,i}}, i = 1,2$. As a result from expression (4.73) we will find at $n = 0$, introducing $M_i^1 = \max\limits_{Q_{t,i}}|u_i^1|$, $z_i^0 = \max\limits_{Q_{t,i}}(u_i^0)^2, z_i^0 \le 1$:

$$M_i^1 \le \frac{B[\exp(t\alpha_i) - 1]}{\alpha_i} = S_i, \quad i = 1,2. \tag{4.74}$$

Hence, at $\alpha_i > 0, i = 1,2$ we find that the top border $M_i^1$ will not surpass 1, if there is inequality $S_i \le 1, i = 1,2$ in Eq. (4.74):

$$t \le \ln\left(\frac{\alpha_i}{B} + 1\right)^{1/\alpha_i}, \quad i = 1,2. \tag{4.75}$$

Therefore choosing an interval $[0,t]$ small enough so that the condition is satisfied from Eq. (4.75), we will have $M_i^1 \le 1, i = 1,2$. Finally we receive definitively $M_i^{n+1} \le S_i z_i^n, i = 1,2$ or

$$\max_{x,t\in\overline{Q_{t,1}}} |w_1^{n+1} - w_1^n| \le S_1 \max_{x,t\in\overline{Q_{t,1}}} |w_1^n - w_1^{n-1}|^2, \tag{4.76}$$

$$\max_{z,t\in\overline{Q_{t,2}}} |w_2^{n+1} - w_2^n| \le S_2 \max_{z,t\in\overline{Q_{t,2}}} |w_2^n - w_2^{n-1}|^2. \tag{4.77}$$

The expressions (4.67) and (4.70) are generally received from the Eqs. (4.29), (4.30), (4.38)–(4.40) with the help of which we have the analytical solution (4.66)–(4.70) of a boundary problem (4.64), fair to all range of definition $\overline{Q_{t,i}}$, $i = 1, 2$, in particular, for $z = 0$. Therefore for Eqs. (4.76) and (4.77), it is possible to write the expression similarly at $z = 0$ $(x = 0)$:

$$\max_{t\in\overline{Q_{t,1}}} |w_0^{n+1} - w_0^n| \le S_0 \max_{t\in\overline{Q_{t,1}}} |w_0^n - w_0^{n-1}|^2, \quad S_0 \le 1. \tag{4.78}$$

Relationships (4.76)–(4.78) show that if convergence of the iterative process for a boundary problem (4.64) or (4.63) according to the inversion formula (4.65) takes place it is quadratic. Thus, with a big enough $n$ each following step doubles a number of correct signs in the given approximation.

## Results of Test Checks

Estimation of accuracy of analytical formulas (4.17), (4.41)–(4.53) are checked practically in the solution of a boundary problem for a two-layer material with various thermal-physical properties in the region $Q_1 = (a_1 < z < 0, -\infty < a_1 < 0)$ and $Q_2 = (0 < z < a_2, 0 < a_2 < \infty)$, $\overline{Q_i} = Q_i + \Gamma$, $\overline{Q_{t,i}} = \overline{Q_i} \times [0 < t \le t_0], i = 1, 2$

$$C_i \frac{\partial T_i}{\partial t} = \frac{\partial}{\partial z}\left[A_i(T_i)\frac{\partial T_i}{\partial z}\right] + B_i(T_i)\frac{\partial T_i}{\partial z} + E_i T_i^k$$
$$+ F_i(z, t), \quad A_i = N_i T_i^m, \quad i = 1, 2, \tag{4.79}$$

$$A_1 \frac{\partial T_1}{\partial z}\bigg|_{z=-0} = A_2 \frac{\partial T_2}{\partial z}\bigg|_{z=+0}, \quad T_1|_{z=-0} = T_2|_{z=+0}, \tag{4.80}$$

$$\left[A_1 \frac{\partial T_1}{\partial z} + S_1(t) T_1\right]_{z=-a_1} = D_1(-a_1, t),$$
$$\left[A_2 \frac{\partial T_2}{\partial z} + S_2(t) T_2\right]_{z=a_2} = D_2(a_2, t), \tag{4.81}$$

$$T_i\mid_{t=0}=T_{H,i}(\gamma_i),\quad \gamma_1=\frac{|z|}{a_1},\quad \gamma_2=\frac{z}{a_2}. \tag{4.82}$$

The exact solution of a problem (4.79)–(4.82) is taken:

$$T_1=\exp(\tau+\gamma_1),\quad T_2=\exp(\tau+\gamma_2),\quad \tau=\frac{t}{c}. \tag{4.83}$$

Then the sources $F_i, D_i, i=1,2$ in the equations above will be

$$F_i=\exp(\tau+\gamma_i)\left\{\frac{C_i}{c}-\left[\frac{B_i}{a_i}+\frac{N_i(m+1)\exp(m(\tau+\gamma_i))}{a_i^2}\right]\right\}$$
$$-\exp[k(\tau+\gamma_i)]E_i,\quad D_i=\exp(\tau+1)$$
$$\times\left\{\frac{N_i}{a_i}\exp[m(\tau+1)]+S_i\right\}.$$

The following basic values of initial data were used: $m=0.5$, $t_0=1$, $A_{H,i}=1, N_i=a_i, c=10, S_i=1, N=M=11, \Delta x_i=a_i/(N-1), i=1,2, \Delta t=0.02$ are numbers of checkouts and steps on space and time while finding integrals in the Eqs. (4.41)–(4.53) by Simpson's formula [21].

The boundary problem (4.79)–(4.82) is solved by means of formulas (4.17), (4.41)–(4.53). The number of iterations was traced [for total expressions of a type (4.51)–(4.53)] on relative change of an error vector:

$$||V_i^n||=\max_{z,t\in\overline{Q_{t,i}}}\left|\frac{w_i^{n+1}-w_i^n}{w_i^{n+1}}\right|,\quad i=1,2.$$

The program is made in the language Fortran-90, calculation was made on Pentium 5 (3.5 GHz, compiler PS 5) with double accuracy. In Table 4.1 the maximum relative error is presented: $\varepsilon=\max(\varepsilon_1,\varepsilon_2)$

$$\varepsilon_i=\frac{|T_i-\tilde{T}_i|100\%}{T_i},\quad i=1,2, \tag{4.84}$$

where $T_i$ is the exact explicit solution (4.83), $\tilde{T}_i, i=1,2$ is the approximate analytical solution on mathematical technology of this paragraph at various values $k$, $B_i, E_i, i=1,2$. In Eq. (4.84) $\varepsilon_i, i=1,2$ answers $a_1=0.95$ m, $a_2=0.9$ m, $C_1=8.61\cdot10^3\,\text{s/m}^2$, $C_2=2\cdot10^5\,\text{s/m}^2$ (a material is of type of copper and a steel [31]).

In Table 4.1 there are the results of test calculations $||V_i^n||\le\delta, i=1,2$ $\delta=0.01$. Thus only 2–3 iterations were enough to achieve this accuracy and time of calculation of any variant is $t_p=1$ s.

As we can see from Table 4.1 calculation on developed mathematical technology has almost the small error $\varepsilon=\max(\varepsilon_1,\varepsilon_2)$ from Eq. (4.84).

**Table 4.1** A dependence of the maximum relative errors at the various values $k, B_i, E_i, i = 1, 2$

| Variant number | Results of calculations | | | | | |
|---|---|---|---|---|---|---|
| | $k$ | $B_1$ | $B_2$ | $E_1$ | $E_2$ | $\varepsilon$, % |
| 1 | 1 | −1 | −1 | −1 | −1 | 5.61 |
| 2 | 2 | −1 | – | −1 | −1 | 5.37 |
| 3 | 2 | −1 | −1 | 1 | 1 | 7.26 |
| 4 | 1 | −1 | −1 | −100 | −10 | 5.93 |
| 5 | 1 | 1 | 1 | 1 | 1 | 13.01 |
| 6 | 1 | −1 | −1 | −10 | 5 | 7.92 |
| 7 | 1 | 1 | 1 | −1 | −1 | 11.29 |

Now we compare the accuracy of analytical solutions (4.51)–(4.56) and (4.60)–(4.62) in solution of a boundary problem (4.57)–(4.59). The following initial data from [2, 45, 46] were taken: $T_\Gamma = 1300\,\text{K}$, $T_H = 293\,\text{K}$, $a_1 = 0.95$ m, $a_2 = 0.9$ m, $\xi_1 = 0.2 \cdot 10^{-4}\,\text{m}^2/\text{s}$, $\xi_2 = 0.17 \cdot 10^{-6}\,\text{m}^2/\text{s}$, $\lambda_1 = 0.0253$ W/(m · K), $\lambda_2 = 0.14$ W/(m · K), $R = 8.31$ J/(mole · K), $D = 1.8 \cdot 10^{-5}\,\text{m}^2/\text{s}$, $M = 0.018$ kg/mole, $k_1 = 2 \cdot 10^5$ m/s, $E_1 = 4.19 \cdot 10^4$ J/mole, $q_1 = 2.5 \cdot 10^6$ J/kg, $k_2 = 2 \cdot 10^8$ m/s, $E_2 = 2.22 \cdot 10^5$ J/mole, $q_2 \rho_w = 10^3$ J/m$^3$, $P_e = 4.4 \cdot 10^3\,\text{N/m}^2$, $\sigma = 5.67 \cdot 10^{-8}$ W/(m$^2$ · K$^4$), $t_0 = 81$ s, $\varepsilon = 0.1$, $c_H = 0.23$, $A = 0.7$.

At present, $t = t_0$ by formulas (4.51)–(4.56), we have $T_w = 797\,\text{K}$, $c_w = 0.217795 \cdot 10^{-3}$, and with the Eqs. (4.60)–(4.62)—$T_w = 877\,\text{K}$, $c_w = 0.217794 \cdot 10^{-3}$. The difference in surface temperature is not more than 9%.

## 4.2 METHOD OF SOLUTION OF THE THREE-DIMENSIONAL CONJUGATE BOUNDARY PROBLEM

### Statement of a Problem and a Method Algorithm

Let us attempt to find the solution of the equation of parabolic type [5, 7] of the adjoining spatial bodies, when in the statement of a corresponding boundary problem, boundary conditions of the fourth general type take place (balance relationships [2, 45])

$$C_1(T_1)\frac{\partial T_1}{\partial t} = \sum_{j=1}^{3} \frac{\partial}{\partial x_j}\left[A_1(T_1)\frac{\partial T_1}{\partial x_j}\right] + B_1(T_1)\frac{\partial T_1}{\partial x_1} + \sum_{j=2}^{3} U_j(T_1)\frac{\partial T_1}{\partial x_j} + Y_1 T_1^k + E_1(x,t), \tag{4.85}$$

$$C_2(T_2)\frac{\partial T_2}{\partial t} = \sum_{j=1}^{3} \frac{\partial}{\partial x_j}\left[A_2(T_2)\frac{\partial T_2}{\partial x_j}\right] + B_2(T_2)\frac{\partial T_2}{\partial x_1}$$

$$+\sum_{j=2}^{3} F_j(T_2)\frac{\partial T_2}{\partial x_j} + Y_2 T_2^k + E_2(x,t) \qquad (4.86)$$

in the region $Q_1 = (-a_1 < x_1 < 0, 0 < x_i < L_i, i = 2, 3)$, $\overline{Q_1} = Q_1 + \Gamma$, $\overline{Q_{t,1}} = \overline{Q_1} \times [0 < t \leq t_0]$ for the Eq. (4.85) and in the region $Q_2 = (0 < x_1 < a_2, 0 < x_i < L_i, i = 2, 3)$, $\overline{Q_2} = Q_2 + \Gamma$, $\overline{Q_{t,2}} = \overline{Q_2} \times [0 < t \leq t_0]$ for the Eq. (4.86), $\Gamma$ is boundary surface of a range of definition $Q_i, i = 1, 2$ at $Y_j = \text{const}, j = 1, 2$ with the initial condition

$$T_i\,|_{t=0} = T_{H,i}(x), \quad i = 1, 2, \quad x = (x_1, x_2, x_3), \qquad (4.87)$$

with a balance boundary condition [2, 45]:

$$A_1 \frac{\partial T_1}{\partial x_1}\bigg|_{x_1=-0} + F(t, x_2, x_3) = A_2 \frac{\partial T_2}{\partial x_1}\bigg|_{x_1=+0},$$

$$T_1|_{x_1=-0} = T_2|_{x_1=+0}, \quad 0 < x_i < L_i, \quad i = 2, 3 \qquad (4.88)$$

and with a boundary condition of the first type

$$T_i|_{\Gamma} = \Psi, \quad i = 1, 2, \qquad (4.89)$$

where $\Psi \neq \text{const}$.

Let's assume everywhere:

1. A problem (4.85)–(4.89) has the unique solution $T_i(x,t), i = 1, 2$, which is continuous in each region $\overline{Q_{t,i}}, i = 1, 2$ and has continuous derivatives $\frac{\partial T_i}{\partial t}, \frac{\partial T_i}{\partial x_j}, \frac{\partial^2 T_i}{\partial x_j^2}, i = 1, 2, j = 1, 2, 3$.
2. The following conditions are satisfied: $A_i(T_i) \geq c_1 > 0$, $C_i(T_i) \geq c_2 > 0$, $c_1, c_2$—constants, $i = 1, 2$; $T_{H,i}$ is a continuous function in $\overline{Q_i}, i = 1, 2$, and $F, A_i, B_i, U_j, F_j, E_i, i = 1, 2, j = 2, 3$ are continuous functions in the closed regions $\overline{Q_{t,i}}, i = 1, 2$.
3. Coefficients $C_i(T_i), B_i(T_i), U_j(T_1), F_j(T_2), i = 1, 2, j = 2, 3$ can be in general dependent on the problem solution $T_i$ [2], the type $A_i(T_i), i = 1, 2$ is defined in the formula (4.91); $\Psi$ is a continuous function on border $\Gamma$ for $0 < t \leq t_0$ having the limited partial derivatives of the first order.

Let's apply the Kirchhoff's transformation [5]

$$v_i = \int_0^{T_i} \frac{A_i(T_i)}{A_{H,i}} dT_i, \quad i = 1, 2, \tag{4.90}$$

where $A_{H,i}, i = 1, 2$ is for example, the heat conductivity factor at the temperature equal to zero.

Further in order to use the inversion formula, $A_i(T_i)$, $i = 1, 2$ in Eq. (4.85) we take

$$A_i(T_i) = N_i T_i^m, \quad m > 0, \ N_i > 0, \ N_i = \text{const}, \ i = 1, 2. \tag{4.91}$$

Then taking into account the relationships [5]:

$$\nabla A_i = \frac{\partial A_i}{\partial T_i} \nabla T_i, \qquad \frac{\partial v_i}{\partial t} = \frac{A_i}{A_{H,i}} \frac{\partial T_i}{\partial t}, \qquad \nabla v_i = \frac{A_i}{A_{H,i}} \nabla T_i, \quad i = 1, 2 \tag{4.92}$$

we receive a boundary problem from Eqs. (4.85)–(4.89)

$$\begin{aligned} \nabla^2 v_1 &= c_1(v_1)\frac{\partial v_1}{\partial t} - \delta_1(v_1)\frac{\partial v_1}{\partial x_1} - \sum_{j=2}^{3} u_j(v_1)\frac{\partial v_1}{\partial x_j} \\ &\quad - \gamma_1(\phi v)_1^{k/s} - e_1(x, t), \\ \nabla^2 v_2 &= c_2(v_2)\frac{\partial v_2}{\partial t} - \delta_2(v_2)\frac{\partial v_2}{\partial x_1} - \sum_{j=2}^{3} f_j(v_2)\frac{\partial v_2}{\partial x_j} \\ &\quad - \gamma_2(\phi v)_2^{k/s} - e_2(x, t), \end{aligned} \tag{4.93}$$

$$v_i|_{t=0} = v_{H,i}(x), \quad v_{H,i} = \frac{T_{H,i}^s}{\phi_i}, \quad \phi_i = \frac{sA_{H,i}}{N_i}, \quad i = 1, 2, \ s = m + 1, \tag{4.94}$$

$$\begin{aligned} &\eta \left.\frac{\partial v_1}{\partial x_1}\right|_{x_1=-0} + M = \left.\frac{\partial v_2}{\partial x_1}\right|_{x_1=+0}, \quad \xi v_1|_{x_1=-0} = v_2|_{x_1=+0}, \\ &\eta = \frac{A_{H,1}}{A_{H,2}}, \quad M = \frac{F(t, x_2, x_3)}{A_{H,2}}, \quad \xi = \eta\frac{N_2}{N_1}, \quad 0 < x_i < L_i, \quad i = 2, 3, \end{aligned} \tag{4.95}$$

$$\begin{aligned} &v_1|_{x_1=-a_1} = S_1, \quad v_2|_{x_1=a_2} = S_2, \quad S_1 = T_1^s(-a_1, x_2, x_3, t)/\phi_1, \\ &S_2 = T_2^s(a_2, x_2, x_3, t)/\phi_2, \end{aligned} \tag{4.96}$$

$$v_1|_{x_2=0} = n_1, \quad v_1|_{x_2=L_2} = N_2, \quad n_1 = T_1^s(x_1, x_3, t)/\phi_1,$$
$$N_2 = T_1^s(x_1, L_2, x_3, t)/\phi_1, \quad v_1|_{x_3=0} = q_1, \quad q_1 = T_1^s(x_1, x_2, t)/\phi_1,$$
$$v_1|_{x_3=L_3} = Q_2, \quad Q_2 = T_1^s(x_1, x_2, L_3, t)/\phi_1, \tag{4.97}$$

$$v_2|_{x_2=0} = p_1, \quad v_2|_{x_2=L_2} = P_2, \quad p_1 = T_2^s(x_1, x_3, t)/\phi_2, \quad v_2|_{x_3=0} = d_1,$$
$$P_2 = T_2^s(x_1, L_2, x_3, t)/\phi_2, \quad d_1 = T_2^s(x_1, x_2, t)/\phi_2,$$
$$v_2|_{x_3=L_3} = D_2, \quad D_2 = T_2^s(x_1, x_2, L_3, t)/\phi_2, \tag{4.98}$$

where $c_i = C_i/A_i, \delta_i = B_i/A_i, \gamma_i = Y_i/A_{H,i}, e_i = E_i/A_{H,i}, i = 1, 2, u_j = U_j/A_1, f_j = F_j/A_2, j = 2, 3$. Thus $T_i, i = 1, 2$ is defined from Eq. (4.90) according to the inversion formula:

$$v_i = T_i^s/\phi_i, \quad T_i = (v\phi)_i^{1/s}, \quad i = 1, 2.$$

Let's note that variation range of independent variables and type of the boundary conditions do not change in relation to Kirchhoff's transformation (4.90) and within the inversion formula the boundary conditions of the first-the fourth types pass into Dirichlet, Neumann, and Newton's conditions and the balance boundary conditions.

Let's apply the locally-one-dimensional scheme of splitting to Eq. (4.93) at the differential level [12]. As a result we have:

$$\frac{\partial^2 v_i^{(1)}}{\partial x_1^2} = c_i \frac{\partial v_i^{(1)}}{\partial t} - \delta_i \frac{\partial v_i^{(1)}}{\partial x_1} - \gamma_i (\phi v^{(1)})_i^{k/s}$$
$$- \sigma_1 e_i(x, t), \quad i = 1, 2, \quad 0 < t < t_*, \tag{4.99}$$

$$\eta \left. \frac{\partial v_1^{(1)}}{\partial x_1} \right|_{x_1=-0} + M = \left. \frac{\partial v_2^{(1)}}{\partial x_1} \right|_{x_1=+0}, \quad \xi v_1^{(1)}|_{x_1=-0} = v_2^{(1)}|_{x_1=+0},$$
$$0 < x_j < L_j, \; j = 2, 3, \quad v_i^{(1)}(0, x) = v_{H,i}(x), \; i = 1, 2, \tag{4.100}$$

$$v_1^{(1)}|_{x_1=-a_1} = S_1(-a_1, t, x_2, x_3), \quad v_2^{(1)}|_{x_1=a_2} = S_2(a_2, t, x_2, x_3); \tag{4.101}$$

$$\frac{\partial^2 v_1^{(2)}}{\partial x_2^2} = c_1 \frac{\partial v_1^{(2)}}{\partial t} - u_2 \frac{\partial v_1^{(2)}}{\partial x_2} - \sigma_2 e_1, \quad -a_1 < x_1 < 0, \quad 0 < t < t_*,$$
$$\frac{\partial^2 v_2^{(2)}}{\partial x_2^2} = c_2 \frac{\partial v_2^{(2)}}{\partial t} - f_2 \frac{\partial v_2^{(2)}}{\partial x_2} - \sigma_2 e_2, \quad 0 < x_1 < a_2, \tag{4.102}$$

$$v_i^{(2)}(0,x) = v_i^{(1)}(t_*,x),\ i=1,2,\quad v_1^{(2)}|_{x_2=0} = n_1(t,x_1,x_3),$$
$$v_1^{(2)}|_{x_2=L_2} = N_2(t,x_1,L_2,x_3),\quad v_2^{(2)}|_{x_2=0} = p_1(t,x_1,x_3),$$
$$v_2^{(2)}|_{x_2=L_2} = P_2(t,x_1,L_2,x_3); \tag{4.103}$$

$$\frac{\partial^2 v_1^{(3)}}{\partial x_3^2} = c_1\frac{\partial v_1^{(3)}}{\partial t} - u_3\frac{\partial v_1^{(3)}}{\partial x_3} - \sigma_3 e_1,\quad -a_1 < x_1 < 0,\quad 0 < t < t_*,$$
$$\frac{\partial^2 v_2^{(3)}}{\partial x_3^2} = c_2\frac{\partial v_2^{(3)}}{\partial t} - f_3\frac{\partial v_2^{(3)}}{\partial x_3} - \sigma_3 e_2,\quad 0 < x_1 < a_2, \tag{4.104}$$

$$v_i^{(3)}(0,x) = v_i^{(2)}(t_*,x),\ i=1,2,\quad v_1^{(3)}|_{x_3=0} = q_1(t,x_1,x_2),$$
$$v_1^{(3)}|_{x_3=L_3} = Q_2(t,x_1,x_2,L_3),\quad v_2^{(3)}|_{x_3=0} = d_1(t,x_1,x_2),$$
$$v_2^{(3)}|_{x_3=L_3} = D_2(t,x_1,x_2,L_3), \tag{4.105}$$

where $\sigma_1+\sigma_2+\sigma_3=1$.

Conversely if to clean the top indexes (1), (2), (3) and term by term to combine the Eqs. (4.99), (4.102), and (4.104) separately on an interval $-a_1 < x_1 < 0$, and then—$0 < x_1 < a_2$, we will have the modified equations from Eq. (4.93).

As $B_i(T_i), i=1,2, U_j(T_1), F_j(T_2), j=2,3$ in Eqs. (4.85) and (4.86), obviously does not depend from $x=(x_1,x_2,x_3)$, we take substitution [37]:

$$v_i^{(1)} = w_i^{(1)}\exp(-x_1 r_i),\ r_i = 0.5\delta_i,\ i=1,2,\quad v_1^{(j)} = w_1^{(j)}\exp(-x_j z_j),$$
$$v_2^{(j)} = w_2^{(j)}\exp(-x_j m_j),\quad z_j = 0.5u_j,\ m_j = 0.5f_j,\ j=2,3, \tag{4.106}$$

to exclude the first partial derivative on space in Eqs. (4.99), (4.102), and (4.104). So the boundary problem (4.99)–(4.105) and the inversion formula will be rewritten

$$\frac{\partial^2 w_i^{(1)}}{\partial x_1^2} = c_i\frac{\partial w_i^{(1)}}{\partial t} - \exp(r_i x_1)\{\gamma_i[(\phi w^{(1)})_i\exp(-r_i x_1)]^{k/s}$$
$$+\sigma_1 e_i(x,t)\} + w_i^{(1)} r_i^2,\quad i=1,2,\quad 0 < t < t_*, \tag{4.107}$$

$$\eta\left(\frac{\partial w_1^{(1)}}{\partial x_1} - r_1 w_1^{(1)}\right)\Bigg|_{x_1=-0} + M = \left(\frac{\partial w_2^{(1)}}{\partial x_1} - r_2 w_2^{(1)}\right)\Bigg|_{x_1=+0},$$
$$\xi w_1^{(1)}|_{x_1=-0} = w_2^{(1)}|_{x_1=+0},\quad w_i^{(1)}(0,x) = w_{H,i}(x),$$
$$w_{H,i} = T_{H,i}^s\exp(r_i x_1)/\phi_i,\quad i=1,2, \tag{4.108}$$

$$w_1^{(1)}|_{x_1=-a_1} = s_1(-a_1, t, x_2, x_3), \quad w_2^{(1)}|_{x_1=a_2} = s_2(a_2, t, x_2, x_3); \quad (4.109)$$

$$\frac{\partial^2 w_1^{(2)}}{\partial x_2^2} = c_1 \frac{\partial w_1^{(2)}}{\partial t} - \sigma_2 \exp(z_2 x_2) e_1 + w_1^{(2)} z_2^2, \ -a_1 < x_1 < 0, \ 0 < t < t_*,$$
$$\frac{\partial^2 w_2^{(2)}}{\partial x_2^2} = c_2 \frac{\partial w_2^{(2)}}{\partial t} - \sigma_2 \exp(m_2 x_2) e_2 + w_2^{(2)} m_2^2,$$
$$0 < x_1 < a_2, \quad 0 < x_3 < L_3, \quad (4.110)$$

$$w_i^{(2)}(0, x) = w_i^{(1)}(t_*, x), \ i = 1, 2, \quad w_1^{(2)}|_{x_2=0} = n_1(t, x_1, x_3),$$
$$w_1^{(2)}|_{x_2=L_2} = n_2(t, x_1, L_2, x_3), \quad w_2^{(2)}|_{x_2=0} = p_1(t, x_1, x_3),$$
$$w_2^{(2)}|_{x_2=L_2} = p_2(t, x_1, L_2, x_3); \quad (4.111)$$

$$\frac{\partial^2 w_1^{(3)}}{\partial x_3^2} = c_1 \frac{\partial w_1^{(3)}}{\partial t} - \sigma_3 \exp(z_3 x_3) e_1 + w_1^{(3)} z_3^2,$$
$$-a_1 < x_1 < 0, \quad 0 < t < t_*,$$
$$\frac{\partial^2 w_2^{(3)}}{\partial x_3^2} = c_2 \frac{\partial w_2^{(3)}}{\partial t} - \sigma_3 \exp(m_3 x_3) e_2 + w_2^{(3)} m_3^2,$$
$$0 < x_1 < a_2, \quad 0 < x_2 < L_2, \quad (4.112)$$

$$w_i^{(3)}(0, x) = w_i^{(2)}(t_*, x), \ i = 1, 2, \quad w_1^{(3)}|_{x_3=0} = q_1(t, x_1, x_2),$$
$$w_1^{(3)}|_{x_3=L_3} = q_2(t, x_1, x_2, L_3), \quad w_2^{(3)}|_{x_3=0} = d_1(t, x_1, x_2),$$
$$w_2^{(3)}|_{x_3=L_3} = d_2(t, x_1, x_2, L_3), \quad (4.113)$$

where $s_1 = \exp(-r_1 a_1) S_1(-a_1, t, x_2, x_3), s_2 = \exp(r_2 a_2) S_2(a_2, t, x_2, x_3)$, $n_2 = \exp(z_2 L_2) N_2(L_2, t, x_1, x_3), p_2 = \exp(m_2 L_2) P_2(L_2, t, x_1, x_3)$, $q_2 = \exp(z_3 L_3) Q_2(L_3, t, x_1, x_2), d_2 = \exp(m_3 L_3) D_2(L_3, t, x_1, x_2)$,

$$T_i^{(1)} = [(w^{(1)}\phi)_i \exp(-r_i x_1)]^{1/s}, \quad i = 1, 2,$$
$$T_0^{(1)} = (w^{(1)}\phi)_0^{1/s}, \phi_0 = s(A_{H,1} + A_{H,2})/(N_1 + N_2),$$
$$T_1^{(2)} = [(w^{(2)}\phi)_1 \exp(-z_2 x_2)]^{1/s},$$
$$T_2^{(2)} = [(w^{(2)}\phi)_2 \exp(-m_2 x_2)]^{1/s},$$
$$T_1(x, t) = T_1^{(3)} = [(w^{(3)}\phi)_1 \exp(-z_3 x_3)]^{1/s},$$
$$T_2(x, t) = T_2^{(3)} = [(w^{(3)}\phi)_2 \exp(-m_3 x_3)]^{1/s}. \quad (4.114)$$

Our purpose is to find the solution to a nonlinear boundary problem if it exists as a limit of sequence of solutions of linear boundary problems. For this purpose, let's take the results of works [7, 15, 16]. Let $w_i^0 = \text{const}, i = 1, 2$ as an initial approximation [as an initial approximation it is better to take $w_{H,i}, i = 1, 2$ from Eq. (4.108)]. We will consider sequence $w_i^n(t, x), i = 1, 2$ defined by a recurrence relationship [7] (the point above corresponds to a partial derivative on time):

$$\frac{\partial^2 w_i^{(1)}}{\partial x_1^2} = f_i + (w_i^{(1)} - w_i^{(0)})\frac{\partial f_i}{\partial w_i^{(0)}} + (\dot{w}_i^{(1)} - \dot{w}_i^{(0)})\frac{\partial f_i}{\partial \dot{w}_i^{(0)}},$$
$$f_i = f_i(w_i^{(0)}, \dot{w}_i^{(0)}), \quad w_i^{(0)} = w_i^n, \quad i = 1, 2, \quad n = 0, 1, 2, \ldots, \tag{4.115}$$

$$\eta \left.\frac{\partial w_1^{(1)}}{\partial x_1}\right|_{x_1=-0} + E w_0^{(1)} + M = \left.\frac{\partial w_2^{(1)}}{\partial x_1}\right|_{x_1=+0}, \quad w_0^{(1)} = w_1^{(1)}|_{x_1=0},$$
$$\xi w_1^{(1)}|_{x_1=-0} = w_2^{(1)}|_{x_1=+0}, \quad E = (\xi r_2 - \eta r_1)|_{x_1=0},$$
$$w_1^{(1)}|_{x_1=-a_1} = s_1, \quad w_2^{(1)}|_{x_1=a_2} = s_2, \quad w_i^{(1)}|_{t=0} = w_{H,i}, \quad i = 1, 2. \tag{4.116}$$

Each function $w_i^{(1)}(t, x)$ in Eqs. (4.115) and (4.116) in a quasi-one-dimensional case is a solution of the linear equation that is a rather important feature of this algorithm. The algorithm comes from an approximation of Newton-Kantorovich's method [8] in functional space.

To reduce further records we will introduce the notations:

$$f_i = c_i \dot{w}_i^{(0)} - \exp(r_i x_1)(\gamma_i Z_i^{k/s} + e_i) + w_i^{(0)} r_i^2, \quad i = 1, 2,$$
$$\Phi_i = \frac{\gamma_i k Z_i^{k/s} A_{H,i}}{Z_i N_i} - r_i^2, \quad r_i = 0.5\frac{B_i}{A_i}, \quad Z_i = (\phi w^{(0)})_i \exp(-r_i x_1),$$
$$R_i^{(1)} = \gamma_i Z_i^{k/s}\left[\exp(r_i x_1) - \frac{w_i^{(0)} k A_{H,i}}{Z_i N_i}\right] + \sigma_1 e_i \exp(r_i x_1),$$
$$c_i(w_i^{(0)}) = \frac{\partial f_i}{\partial \dot{w}_i^{(0)}}, \quad \Phi_i(w_i) = -\frac{\partial f_i}{\partial w_i^{(0)}}, \quad h_i^{(1)} = c_i \dot{w}_i^{(1)} - R_i^{(1)}(w_i^{(0)}),$$
$$i = 1, 2, \quad h_1^{(j)} = c_1 \dot{w}_1^{(j)} - D_j, \quad \Psi_j = -z_j^2, \quad h_2^{(j)} = c_2 \dot{w}_2^{(j)} - S_j,$$
$$\Theta_j = -m_j^2, \quad z_j = 0.5 U_j / A_1, \quad m_j = 0.5 F_j / A_2,$$
$$D_j = \sigma_j e_1 \exp(z_j x_j), \quad S_j = \sigma_j e_2 \exp(m_j x_j), \quad j = 2, 3. \tag{4.117}$$

Let's take the solution of a boundary problem (4.115) and (4.116) on coordinate to direction $x_1$, using for this purpose the Eqs. (4.107)–(4.109) and notations (4.117) with an index (1) above:

$$\frac{\partial^2 w_i^{(1)}}{\partial x_1^2} + w_i^{(1)}\Phi_i = h_i^{(1)}(w_i^{(0)}, \dot{w}_i^{(1)}, x, t), \quad 0 < t < t_*,$$

$$i = 1, 2, \quad n = 0, 1, 2, \ldots, \tag{4.118}$$

$$\eta \left.\frac{\partial w_1^{(1)}}{\partial x_1}\right|_{x_1=-0} + Ew_0^{(1)} + M = \left.\frac{\partial w_2^{(1)}}{\partial x_1}\right|_{x_1=+0}, \quad w_0^{(1)} = w_1^{(1)}|_{x_1=0},$$

$$\xi w_1^{(1)}|_{x_1=-0} = w_2^{(1)}|_{x_1=+0}, w_i^{(1)}|_{t=0} = w_{H,i}, \quad i = 1, 2,$$

$$w_1^{(1)}|_{x_1=-a_1} = s_1, \quad w_2^{(1)}|_{x_1=a_2} = s_2. \tag{4.119}$$

For a quasi-one-dimensional problem (4.118) and (4.119) we use the algorithm (4.27)–(4.114), (4.54)–(4.56) from the previous paragraph of this chapter. As well as in Section 4.1 by replacement of the variable $x_1 = -z$ we will transform the region $-\infty < x_1 \le 0$ into the region $0 \le z < \infty$.

We assume that the desired solution $w_1^{(1)}$, and its derivatives entering into the Eq. (4.118) satisfy the Laplace integral transformation conditions on $z$, and growth degree on $z$ functions $w_1^{(1)}$ and its derivatives does not depend on $x_2$, $x_3$. Multiplying both parts of the Eqs. (4.118) at $i = 1$ on $\exp(-pz)$ and integrating on $z$ from 0 to $\infty$, we will find [6] introducing capital letters for images $W_i^{(1)}, H_i^{(1)}, i = 1, 2$ and $g_1 = w_1^{(1)}(t, 0, x_2, x_3)$, $\frac{\partial g_1}{\partial z} = \frac{\partial w_1^{(1)}(t,0,x_2,x_3)}{\partial z}$:

$$p^2 W_1^{(1)}(t, p, x_2, x_3) - pg_1 - \frac{\partial g_1}{\partial z} + \Phi_1 W_1^{(1)}(t, p, x_2, x_3)$$

$$= H_1^{(1)}(t, p, x_2, x_3), \quad 0 < x_2 < L_2, \quad 0 < x_3 < L_3,$$

$$W_1^{(1)} = \frac{pg_1}{p^2 + b_1^2} + \frac{b_1 \frac{\partial g_1}{\partial z}}{b_1(p^2 + b_1^2)} + \frac{H_1^{(1)} b_1}{b_1(p^2 + b_1^2)}, \quad b_1 = \sqrt{\Phi_1}. \tag{4.120}$$

Using the return Laplace integral transformation [6]: $L^{-1}[p/(p^2 + b_1^2)] = \cos(b_1 z)$ at $b_1^2 > 0$, $L^{-1}[p/(p^2 - b_1^2)] = \cosh(b_1 z)$ at $b_1^2 < 0$;

$L^{-1}[H_1(p)/p] = \int_0^z h_1(y)\,dy$, we restore the original for $w_1^{(1)}(t, z, x_2, x_3)$ from Eq. (4.120) [6]

$$w_1^{(1)}(t, z, x_2, x_3) = g_1 u_1(z) + u_2(z)\frac{\partial g_1}{\partial z} + \int_0^z u_2(z-y)h_1^{(1)}(y)\,dy, \quad (4.121)$$

where $u_1(z) = \cos(b_1 z), u_2(z-y) = b_1^{-1}\sin[b_1(z-y)]u_2(z) = b_1^{-1} \times \sin(b_1 z)$, at $b_1^2 = \Phi_1 > 0$ from Eq. (4.117); $u_1(z) = \cosh(b_1 z),\ u_2(z) == b_1^{-1}\sinh(b_1 z), u_2(z-y) = b_1^{-1}\sinh[b_1(z-y)]$ at $b_1^2 < 0$.

Similarly for $i = 2$ according to the algorithm (4.117), (4.120), and (4.121) in an interval $0 < x_1 < \infty$ for the Eq. (4.118) we find at $b_2 = \sqrt{\Phi_2}$, $g_2 = w_2^{(1)}(t, 0, x_2, x_3)$, $\frac{\partial w_2^{(1)}(t,0,x_2,x_3)}{\partial x_1} = \frac{\partial g_2}{\partial x_1}$:

$$w_2^{(1)}(t, x) = g_2 u_1(x_1) + u_2(x_1)\frac{\partial g_2}{\partial x_1} + \int_0^{x_1} u_2(x_1-y)h_2^{(1)}(y)\,dy. \quad (4.122)$$

Substituting in the Eq. (4.122) instead of $g_2, \frac{\partial g_2}{\partial x_1}$ their expressions from balance conditions (4.119) we have

$$w_2^{(1)} = g_1[\xi u_1(x_1) + u_2(x_1)E] + u_2(x_1)\eta\frac{\partial g_1}{\partial z} + u_2(x_1)M$$
$$+ \int_0^{x_1} u_2(x_1-y)h_2^{(1)}(y)\,dy. \quad (4.123)$$

Function $g_1$ and derivative $\frac{\partial g_1}{\partial z}$ in the Eqs. (4.121) and (4.123) can be found with the help of boundary conditions of the first type from Eq. (4.119)

$$s_1 = g_1 u_1(a_1) + u_2(a_1)\frac{\partial g_1}{\partial z} + \int_0^{a_1} u_2(a_1-y)h_1^{(1)}(y)\,dy,$$

$$s_2 = g_1[\xi u_1(a_2) + u_2(a_2)E] + u_2(a_2)\left(\eta\frac{\partial g_1}{\partial z} + M\right)$$
$$+ \int_0^{a_2} u_2(a_2-y)h_2^{(1)}(y)\,dy,$$

$$g_1 u_1(a_1) + \frac{\partial g_1}{\partial z}u_2(a_1) = s_1 - \int_0^{a_2} u_2(a_2-y)h_2^{(1)}(y)\,dy,$$
$$g_1[\xi u_1(a_2) + u_2(a_2)E] + \frac{\partial g_1}{\partial z}u_2(a_2)\eta$$
$$= s_2 - u_2(a_2)M - \int_0^{a_2} u_2(a_2-y)h_2^{(1)}(y)\,dy,$$

$$g_1 b_1 + \frac{\partial g_1}{\partial z} b_2 = b_3, \quad g_1 c_1 + \frac{\partial g_1}{\partial z} c_2 = c_3, \tag{4.124}$$

$$\begin{aligned} &b_1 = u_1(a_1), \quad b_2 = u_2(a_1), \quad b_3 = s_1 - \int_0^{a_1} u_2(a_1 - y) h_1^{(1)}(y)\, dy, \\ &c_1 = \xi u_1(a_2) + u_2(a_2)E, \quad c_2 = \eta u_2(a_2), \quad c_3 = s_2 - u_2(a_2)M \\ &\quad - \int_0^{a_2} u_2(a_2 - y) h_2^{(1)}(y)\, dy. \end{aligned} \tag{4.125}$$

Finally, from Eq. (4.124) we will have, taking into account formulas (4.125)

$$\begin{aligned} &g_1 = \frac{b_3 c_2 - b_2 c_3}{\Delta}, \quad \frac{\partial g_1}{\partial z} = \frac{b_1 c_3 - b_3 c_1}{\Delta}, \quad \Delta = b_1 c_2 - b_2 c_1, \\ &\Delta = u_1(a_1) u_2(a_2)\eta - u_2(a_1)[\xi u_1(a_2) + u_2(a_2)E], \end{aligned}$$

$$\begin{aligned} g_1 = \Delta^{-1} \Bigg\{ &\left[ s_1 - \int_0^{a_1} u_2(a_1 - y) h_1^{(1)}(y)\, dy \right] \eta u_2(a_2) \\ &- u_2(a_1) \left[ s_2 - u_2(a_2)M - \int_0^{a_2} u_2(a_2 - y) h_2^{(1)}(y)\, dy \right] \Bigg\}, \end{aligned}$$

$$\begin{aligned} \frac{\partial g_1}{\partial z} = \Delta^{-1} \Bigg\{ &u_1(a_1) \left[ s_2 - u_2(a_2)M - \int_0^{a_2} u_2(a_2 - y) h_2^{(1)}(y)\, dy \right] \\ &- [\xi u_1(a_2) + u_2(a_2)E] \left[ s_1 - \int_0^{a_1} u_2(a_1 - y) h_1^{(1)}(y)\, dy \right] \Bigg\}. \end{aligned} \tag{4.126}$$

Let's substitute $g_1$, $\frac{\partial g_1}{\partial z}$ from Eq. (4.126) in the Eqs. (4.121) and (4.123) and introduce notations $w_0^{(1)} = w_0^{(1)}(t, 0, x_2, x_3)$ so we have

$$\begin{aligned} &\dot{w}_0^{(1)} + U_0^{(1)} w_0^{(1)} = X_0^{(1)}, \quad \dot{w}_i^{(1)} + U_i^{(1)} w_i^{(1)} = X_i^{(1)}, \\ &i = 1, 2, \quad 0 < x_j < L_j, \quad j = 2, 3, \end{aligned} \tag{4.127}$$

$$\begin{aligned} &X_0^{(1)} = (K_0^{(1)} + Y_0^{(1)} R_0^{(1)})/(C_0 Y_0^{(1)}), \quad C_0 = 0.5(C_1 + C_2)_0 / A_0, \\ &R_0 = 0.5(R_1^{(1)} + R_2^{(1)})_0, \quad U_0^{(1)} = 1/(Y_0^{(1)} C_0), \\ &K_0^{(1)} = \Delta^{-1} \{ s_1 \eta u_2(a_2) - [s_2 - u_2(a_2)M] u_2(a_1) \}, \\ &Y_0^{(1)} = \Delta^{-1} \left[ \eta u_2(a_2) \int_0^{a_1} u_2(a_1 - y)\, dy - u_2(a_1) \int_0^{a_2} u_2(a_2 - y)\, dy \right], \\ &\Delta = \eta u_1(a_1) u_2(a_2) - u_2(a_1)[\xi u_1(a_2) + E u_2(a_2)]; \end{aligned}$$

$$X_i^{(1)} = (K_i^{(1)} + F_i^{(1)} + V_i^{(1)})/(c_i Y_i^{(1)}), \quad i = 1, 2,$$
$$K_1^{(1)}(t, z, x_2, x_3) = \Delta^{-1} u_1(z)\{\eta s_1 u_2(a_2) + [u_2(a_2)M - s_2]u_2(a_1)\}$$
$$+ \frac{\eta u_2(z)}{\Delta}\{u_1(a_1)[s_2 - u_2(a_2)M]$$
$$- s_1[\xi u_1(a_2) + E u_2(a_2)]\}, \quad 0 < x_j < L_j, \quad j = 2, 3, \tag{4.128}$$

$$w_1^{(1)} = -\int_0^{a_1} G_1^{(1)}(z, y) h_1^{(1)}(y)\, dy + V_1^{(1)}(h_2^{(1)}) + K_1^{(1)},$$
$$F_1^{(1)} = \int_0^{a_1} G_1^{(1)}(z, y) R_1^{(1)}(y)\, dy, \quad U_1^{(1)} = 1/(c_1 Y_1^{(1)}), \quad c_1 = C_1/A_1,$$
$$h_1^{(1)} = [V_1^{(1)}(h_2^{(1)}) + K_1^{(1)} - w_1^{(1)}]/Y_1^{(1)}, \tag{4.129}$$

$$G_1^{(1)}(z, y) = \begin{cases} P_1^{(1)} u_2(a_1 - y) - u_2(z - y), & 0 \le y \le z, \\ P_1^{(1)} u_2(a_1 - y), & z \le y \le a_1, \end{cases} \tag{4.130}$$

$$P_1^{(1)}(t, z, x_2, x_3) = \Delta^{-1}\{\eta u_1(z) u_2(a_2) - u_2(z)[\xi u_1(a_2) + E u_2(a_2)]\},$$
$$V_1^{(1)}(h_2^{(1)}) = \Delta^{-1}[u_1(z) u_2(a_1) - u_2(z) u_1(a_1)] \int_0^{a_2} u_2(a_2$$
$$- y) h_2^{(1)}\, dy, \quad Y_1^{(1)} = \int_0^{a_1} G_1^{(1)}(z, y)\, dy, \quad 0 < x_j < L_j, \quad j = 2, 3; \tag{4.131}$$

$$w_2^{(1)} = -\int_0^{a_2} G_2^{(1)}(x_1, y) h_2^{(1)}(y)\, dy + V_2^{(1)}(h_1^{(1)}) + K_2^{(1)}(t, x),$$
$$K_2^{(1)}(t, x) = \frac{\xi u_1(x_1) + E u_2(x_1)}{\Delta}\{\eta s_1 u_2(a_2)$$
$$+ [u_2(a_2)M - s_2]u_2(a_1)\} + \frac{\eta u_2(x_1)}{\Delta}\{u_1(a_1)[s_2 - u_2(a_2)M] - s_1[\xi u_1(a_2)$$
$$+ u_2(a_2)E]\} + u_2(x_1)M, \quad U_2^{(1)} = 1/(c_2 Y_2^{(1)}), \quad c_2 = C_2/A_2,$$
$$h_2^{(1)} = [V_2^{(1)}(h_1^{(1)}) + K_2^{(1)}(t, x) - w_2^{(1)}]/Y_2^{(1)}, \tag{4.132}$$

$$G_2^{(1)}(x_1, y) = \begin{cases} P_2^{(1)}(t, x) u_2(a_2 - y) - u_2(x_1 - y), & 0 \le y \le x_1, \\ P_2^{(1)}(t, x) u_2(a_2 - y), & x_1 \le y \le a_2, \end{cases} \tag{4.133}$$

$$P_2^{(1)}(t, x) = \Delta^{-1}\{\eta u_2(x_1) u_1(a_1) - [\xi u_1(x_1) + E u_2(x_1)] u_2(a_1)\},$$

$$V_2^{(1)}(h_1^{(1)}) = \Delta^{-1}\eta\{u_2(x_1)[\xi u_1(a_2) + u_2(a_2)E] - u_2(a_2)[\xi u_1(x_1) + u_2(x_1)E]\} \int_0^{a_1} u_2(a_1 - y)h_1^{(1)}\, dy,$$

$$Y_2^{(1)} = \int_0^{a_2} G_2^{(1)}(x_1, y)\, dy, \quad F_2^{(1)} = \int_0^{a_2} G_2^{(1)}(x_1, y)R_2^{(1)}(y)\, dy,$$

$$u_1(x_1) = \cos(b_2 x_1), \quad u_2(x_1) = b_2^{-1}\sin(b_2 x_1), \quad u_2(x_1 - y) = b_2^{-1}\sin[b_2(x_1 - y)] \quad \text{at} \quad b_2^2 = \Phi_2 > 0 \quad \text{from} \quad \text{Eq. (4.117)};$$
$$u_1(x_1) = ch(b_2 x_1), \quad u_2(x_1) = b_2^{-1} sh(b_2 x_1), \quad u_2(x_1 - y) = b_2^{-1} sh[b_2(x_1 - y)] \quad \text{at} \quad b_2^2 < 0. \tag{4.134}$$

As a result the problem solution (4.107)–(4.109) will be [33]:

$$w_0^{(1)}(0, x_2, x_3, t_*) = \left[ w_{H,0} + \int_0^{t_*} X_0^{(1)}(w_0^n, \tau) \exp(\tau U_0^{(1)})\, d\tau \right] \times \exp(-t_* U_0^{(1)}), \quad 0 < x_j < L_j, \quad j = 2, 3, \tag{4.135}$$

$$w_i^{(1)}(x, t_*) = \left[ w_{H,i} + \int_0^{t_*} X_i^{(1)}(w_i^n, x, \tau) \exp(\tau U_i^{(1)})\, d\tau \right] \times \exp(-t_* U_i^{(1)}), \quad -a_1 < x_1 < a_2, \quad x_1 \neq 0,$$
$$i = 1, 2, \quad 0 < x_j < L_j, \quad j = 2, 3. \tag{4.136}$$

Let's apply the Laplace integral transformation to the differential equation (4.110) excluding derivative on $x_2$ and replacing it with linear expression in relation to the required function. Then we consider the functions for which the Laplace integral transformation converges absolutely. The valid part of complex number $p = \xi + i\eta, i = \sqrt{-1}$ is considered positive, etc. Re $p > 0$. Let's introduce capital letters for images $W_i^{(2)}, H_i^{(2)}$, $i = 1, 2$. We have:

$$W_i^{(2)}(t, x_1, p, x_3) = \int_0^{\infty} \exp(-px_2) w_i^{(2)}(t, x)\, dx_2,$$

$$w_i^{(2)}(t, x) = L^{-1}[W_i^{(2)}(t, x_1, p, x_3)], \quad h_i^{(2)}(t, x) = L^{-1}[H_i^{(2)}(t, x_1, p, x_3)],$$
$i = 1, 2.$

We assume that the desired solution $w_i^{(2)}(t, x), i = 1, 2$ and its derivatives entering into the Eq. (4.110) satisfy Laplace integral transformation conditions on $x_2$, and growth degree on $x_2$ functions $w_i^{(2)}(t, x), i = 1, 2$ and its derivatives do not depend on $z$, $x_1$, $x_3$. Multiplying both parts of the Eqs. (4.110) on $\exp(-px_2)$ and integrating on $x_2$ from 0 to $\infty$, we will receive [6] at $g_2 = w_1^{(2)}(t,z,0,x_3)$, $(\partial w_1^{(2)}/\partial x_2)|_{x_2=0} = \partial g_2/\partial x_2$ in the field of $0 < z < a_1$:

$$p^2 W_1^{(2)}(t,z,p,x_3) - pg_2(t,z,0,x_3) - \frac{\partial g_2(t,z,0,x_3)}{\partial x_2}$$
$$+ \Psi_2 W_1^{(2)}(t,z,p,x_3) = H_1^{(2)}(t,z,p,x_3), \quad 0 < z < a_1, \quad 0 < x_3 < L_3,$$
$$W_1^{(2)} = \frac{pg_2}{p^2+\beta_2^2} + \frac{\beta_2 \partial g_2/\partial x_2}{\beta_2(p^2+\beta_2^2)} + \frac{H_1^{(2)}\beta_2}{\beta_2(p^2+\beta_2^2)}, \quad \beta_2 = \sqrt{\Psi_2}. \quad (4.137)$$

Using the Laplace integral transformation [6], formulas (4.117), algorithms from Sections 3.2 and 4.1 of both Chapters 3 and 4, we have in the region $\overline{Q_{t,i}}$, $i = 1, 2$:

$$w_0^{(2)}(0,x_2,x_3,t_*) = \left[ w_0^{(1)}(t_*,0,x_2,x_3) + \int_0^{t_*} X_0^{(2)}(w_0^{(1)},\tau)\exp(\tau U_0^{(2)}) \right.$$
$$\left. \times\, d\tau \right] \exp(-t_* U_0^{(2)}), \quad 0 < x_2 < L_2, \ 0 < x_3 < L_3, \tag{4.138}$$

$$X_0^{(2)} = \left[ K_0^{(2)} + \int_0^{L_2} G_1^{(2)}(x_2,y) D_{2,0}(0,y,x_3,t)\,dy \right] / Y_0^{(2)},$$

$$Y_0^{(2)} = C_0 \int_0^{L_2} \frac{G_1^{(2)}(x_2,y)}{A_{1,0}}\,dy, \quad C_0 = (C_1 + C_2)_0/2, \quad U_0^{(2)} = 1/Y_0^{(2)},$$

$$K_0^{(2)}(0,x_2,x_3,t) = n_1 u_1(x_2) + u_2(x_2)[n_2 - n_1 u_1(L_2)]/u_2(L_2),$$

$$G_1^{(2)}(x_2,y) = \begin{cases} u_2(x_2)u_2(L_2-y)/u_2(L_2) - u_2(x_2-y), & 0 \le y \le x_2, \\ u_2(x_2)u_2(L_2-y)/u_2(L_2), & x_2 \le y \le L_2, \end{cases} \tag{4.139}$$

$$u_1(x_2) = \cos(\beta_2 x_2), \quad u_2(x_2) = \beta_2^{-1}\sin(\beta_2 x_2), \quad u_2(x_2-y)$$
$$= \beta_2^{-1}\sin[\beta_2(x_2-y)] \quad \text{at} \quad \beta_2^2 = \Psi_2 > 0 \quad \text{from Eq. (4.117)};$$
$$u_1(x_2) = ch(\beta_2 x_2), \quad u_2(x_2) = \beta_2^{-1} sh(\beta_2 x_2), \quad u_2(x_2-y)$$
$$= \beta_2^{-1} sh[\beta_2(x_2-y)] \quad \text{at} \quad \beta_2^2 < 0.$$

$$w_i^{(2)}(x, t_*) = \left[ w_i^{(1)}(t_*(X)) + \int_0^{t_*} X_i^{(2)}(w_i^{(1)}, x, \tau) \exp(\tau U_i^{(2)})\, d\tau \right]$$
$$\times \exp(-t_* U_i^{(2)}), \quad i = 1, \quad 0 < z < a_1,\ 0 < x_3 < L_3,$$
$$i = 2, \quad 0 < x_1 < a_2, \quad 0 < x_3 < L_3, \tag{4.140}$$

$$X_1^{(2)} = \left[ K_1^{(2)} + \int_0^{L_2} G_1^{(2)}(x_2, y) D_2(z, y, x_3, t)\, dy \right] / Y_1^{(2)},$$
$$X_2^{(2)} = \left[ K_2^{(2)} + \int_0^{L_2} G_2^{(2)}(x_2, y) S_2(x_1, y, x_3, t)\, dy \right] / Y_2^{(2)},$$

$$K_1^{(2)}(t, x) = n_1 u_1(x_2) + u_2(x_2)[n_2 - n_1 u_1(L_2)]/u_2(L_2),$$
$$K_2^{(2)}(t, x) = p_1 v_1(x_2) + v_2(x_2)[p_2 - p_1 v_1(L_2)]/v_2(L_2),$$

$$Y_1^{(2)} = C_1 \int_0^{L_2} \frac{G_1^{(2)}(x_2, y)}{A_1}\, dy, \quad Y_2^{(2)} = C_2 \int_0^{L_2} \frac{G_2^{(2)}(x_2, y)}{A_2}\, dy,$$
$$U_i^{(2)} = 1/Y_i^{(2)}, \quad i = 1, 2,$$

$$G_2^{(2)}(x_2, y) = \begin{cases} v_2(x_2) v_2(L_2 - y)/v_2(L_2) - v_2(x_2 - y), & 0 \le y \le x_2, \\ v_2(x_2) v_2(L_2 - y)/v_2(L_2), & x_2 \le y \le L_2, \end{cases} \tag{4.141}$$

$$v_1(x_2) = \cos(\gamma_2 x_2), \quad v_2(x_2) = \gamma_2^{-1} \sin(\gamma_2 x_2), \quad v_2(x_2 - y)$$
$$= \gamma_2^{-1} \sin[\gamma_2(x_2 - y)] \quad \text{at} \quad \gamma_2^2 = \Theta_2 > 0 \quad \text{from} \quad \text{Eq. (4.117)};$$
$$v_1(x_2) = ch(\gamma_2 x_2), \quad v_2(x_2) = \gamma_2^{-1} sh(\gamma_2 x_2), \quad v_2(x_2 - y)$$
$$= \gamma_2^{-1} sh[\gamma_2(x_2 - y)] \quad \text{at} \quad \gamma_2^2 < 0.$$

Similarly in the third coordinate direction $x_3$ we receive the formulas similar to Eqs. (4.137)–(4.141). The solution of a boundary problem finally Eqs. (4.107)–(4.113) looks like:

$$w_0^{(3)}(x_2, x_3, t_*) = \exp(-t_* U_0^{(3)}) \left[ w_0^{(2)}(x_2, x_3, t_*) + \int_0^{t_*} X_0^{(3)}(w_0^{(2)}, \tau) \right.$$
$$\left. \times \exp(\tau U_0^{(3)})\, d\tau \right], \quad 0 < x_j < L_j,\ j = 2, 3,\ n = 0, 1, 2, \ldots, \tag{4.142}$$

$$X_0^{(3)} = \left[ K_0^{(3)}(0, x_2, x_3, t) + \int_0^{L_3} G_1^{(3)}(x_3, y) D_{3,0}(0, x_2, y, t)\, dy \right] / Y_0^{(3)},$$
$$K_0^{(3)}(0, x_2, x_3, t) = d_1 u_1(x_3) + u_2(x_3)[d_2 - d_1 u_1(L_3)]/u_2(L_3),$$
$$Y_0^{(3)} = C_0 \int_0^{L_3} \frac{G_1^{(3)}(x_3, y)}{A_{1,0}}\, dy, \quad U_0^{(3)} = 1/Y_0^{(3)},$$

$$G_1^{(3)}(x_3, y) = \begin{cases} u_2(x_3)u_2(L_3 - y)/u_2(L_3) - u_2(x_3 - y), & 0 \le y \le x_3, \\ u_2(x_3)u_2(L_3 - y)/u_2(L_3), & x_3 \le y \le L_3, \end{cases} \tag{4.143}$$

$$u_1(x_3) = \cos(\beta_3 x_3), \quad u_2(x_3) = \beta_3^{-1}\sin(\beta_3 x_3), \quad u_2(x_3 - y)$$
$$= \beta_3^{-1}\sin[\beta_3(x_3 - y)] \text{ at } \beta_3^2 = \Psi_3 > 0 \text{ from Eqs. (4.117) and (4.137);}$$
$$u_1(x_3) = ch(\beta_3 x_3), \quad u_2(x_3) = \beta_3^{-1} sh(\beta_3 x_3), \quad u_2(x_3 - y)$$
$$= \beta_3^{-1} sh[\beta_3(x_3 - y)] \quad \text{at} \quad \beta_3^2 < 0,$$

$$w_i^{(3)}(x, t_*) = \left[ w_i^{(2)}(x, t_*) + \int_0^{t_*} X_i^{(3)}(w_i^{(2)}, x, \tau) \exp(\tau U_i^{(3)})\, d\tau \right]$$
$$\times \exp(-t_* U_i^{(3)}), \quad i = 1, \quad 0 < z < a_1,\ 0 < x_2 < L_2,$$
$$i = 2, \quad 0 < x_1 < a_2, \quad 0 < x_2 < L_2, \quad n = 0, 1, 2, \ldots, \tag{4.144}$$

$$X_1^{(3)} = \left[ K_1^{(3)} + \int_0^{L_3} G_1^{(3)}(x_3, y) D_3(z, x_2, y, t)\, dy \right] / Y_1^{(3)},$$
$$X_2^{(3)} = \left[ K_2^{(3)} + \int_0^{L_3} G_2^{(3)}(x_3, y) S_3(x_1, x_2, y, t)\, dy \right] / Y_2^{(3)},$$

$$K_1^{(3)}(t, x) = q_1 u_1(x_3) + u_2(x_3)[q_2 - q_1 u_1(L_3)]/u_2(L_3),$$
$$K_2^{(3)}(t, x) = d_1 v_1(x_3) + v_2(x_3)[d_2 - d_1 v_1(L_3)]/v_2(L_3),$$

$$Y_1^{(3)} = C_1 \int_0^{L_3} \frac{G_1^{(3)}(x_3, y)}{A_1}\, dy, \quad Y_2^{(3)} = C_2 \int_0^{L_3} \frac{G_2^{(3)}(x_3, y)}{A_2}\, dy,$$
$$U_i^{(3)} = 1/Y_i^{(3)}, \quad i = 1, 2,$$

$$G_2^{(3)}(x_3, y) = \begin{cases} v_2(x_3)v_2(L_3 - y)/v_2(L_3) - v_2(x_3 - y), & 0 \le y \le x_3, \\ v_2(x_3)v_2(L_3 - y)/v_2(L_3), & x_3 \le y \le L_3, \end{cases} \tag{4.145}$$

$$\begin{aligned} v_1(x_3) &= \cos(\gamma_3 x_3), \quad v_2(x_3) = \gamma_3^{-1}\sin(\gamma_3 x_3), \quad v_2(x_3 - y) \\ &= \gamma_3^{-1}\sin[\gamma_3(x_3 - y)] \quad \text{at} \quad \gamma_3^2 = \Theta_3 > 0 \quad \text{from Eq. (4.117);} \\ v_1(x_3) &= ch(\gamma_3 x_3), \quad v_2(x_3) = \gamma_3^{-1} sh(\gamma_3 x_3), \quad v_2(x_3 - y) \\ &= \gamma_3^{-1} sh[\gamma_3(x_3 - y)] \quad \text{at} \quad \gamma_3^2 < 0. \end{aligned}$$

As we can see, the intermediate values are excluded from the algorithm $w_i^{(j)}, T_i^{(j)}, j = 1, 2, 3, i = 0, 1, 2$ from the Eqs. (4.114), (4.120)–(4.136), (4.138)–(4.141) and final formulas (4.142)–(4.145) are formed for $w_i^{(3)} = w_i^{n+1}, i = 0, 1, 2$, then from the inversion formula (4.114)—expressions $T_i^{(3)} = T_i^{n+1}, i = 0, 1, 2$, and then the iterative process $n = 0, 1, 2\ldots$ is included. As a result we receive the solution of a nonlinear initial boundary problem (4.85)–(4.89).

Let's take the analytical solution of a problem for two adjoining semi-infinite (on an axis $x_1$) bodies made of different materials at various initial temperatures, the heat conductivity equations for which look like

$$\frac{\partial T_1}{\partial t} = \sum_{j=1}^{3} \xi_1 \frac{\partial^2 T_1}{\partial x_j^2}, \quad -\infty < x_1 < 0, \quad 0 < x_i < L_i, \quad i = 2, 3, \tag{4.146}$$

$$\frac{\partial T_2}{\partial t} = \sum_{j=1}^{3} \xi_2 \frac{\partial^2 T_2}{\partial x_j^2}, \quad 0 < x_1 < \infty, \quad 0 < x_i < L_i, \quad i = 2, 3, \tag{4.147}$$

where $\xi_i, i = 1, 2$ are coefficients of thermal diffusivity of materials 1 and 2 accordingly.

On the material interface the conditions of equality of thermal streams and temperatures are satisfied

$$\lambda_1 \left.\frac{\partial T_1}{\partial x_1}\right|_{x_1=-0} = \lambda_2 \left.\frac{\partial T_2}{\partial x_1}\right|_{x_1=+0}, \quad T_1|_{x_1=-0} = T_2|_{x_1=+0}, \tag{4.148}$$

where $\lambda_i, i = 1, 2$ are coefficients of heat conductivity of materials 1 and 2.

In the left part of the region $x_1 \leq 0$ (material 1) in its center: $u_1 \leq x_2 \leq u_2$, $v_1 \leq x_3 \leq v_2$ $(u_1 = v_1,\ u_2 = v_2)$ on an axis $x_1$ there is a heated piece with initial temperature $T_{1*} = 800\,\text{K}$. In other parts of material 1: $0 < x_2 < u_1$, $u_2 < x_2 < L_2$, $0 < x_3 < v_1$, $v_2 < x_3 < L_3$ and in material 2 the temperature in the initial time moment is equal to $T_{i,H} = 293\,\text{K}, i = 1, 2$ everywhere:

$$
\begin{aligned}
&T_1(0,x) = T_{1*}, \quad x_1 \leq 0, \quad u_1 \leq x_2 \leq u_2, \quad v_1 \leq x_3 \leq v_2, \\
&T_1(0,x) = T_{1,H}, \quad x_1 \leq 0, \quad 0 < x_2 < u_1, \quad u_2 < x_2 < L_2, \\
&\qquad 0 < x_3 < v_1, \quad v_2 < x_3 < L_3, \\
&T_2(0,x) = T_{2,H}, \quad x_1 > 0, \quad 0 < x_j < L_j,\ j = 2, 3.
\end{aligned}
\tag{4.149}
$$

For simplicity of reception of analytical formulas and further analysis of the solution to a problem, we consider boundary conditions of the first type:

$$
\begin{aligned}
&T_1|_{x_1 \to -\infty} = T_{1*}, \quad u_1 \leq x_2 \leq u_2, \quad v_1 \leq x_3 \leq v_2, \\
&T_1|_{x_1 \to -\infty} = T_{1,H}, \quad 0 < x_2 < u_1, \quad u_2 < x_2 < L_2, \\
&\qquad 0 < x_3 < v_1, \quad v_2 < x_3 < L_3,
\end{aligned}
\tag{4.150}
$$

$$
\begin{aligned}
&T_2|_{x_1 \to \infty} = T_{2,H}, \quad T_i|_{x_j = 0} = T_{i,H}, \\
&T_i|_{x_j = L_j} = T_{i,H}, \quad i = 1, 2, \quad j = 2, 3.
\end{aligned}
\tag{4.151}
$$

As a result of application of algorithm of the locally-one-dimensional scheme of splitting [12, 13] at differential level to a boundary problem (4.146)–(4.151) we have

$$
\frac{\partial T_i^{(1)}}{\partial t} = \xi_i \frac{\partial^2 T_i^{(1)}}{\partial x_1^2}, \quad i = 1, 2,\ 0 < t < t_*,
\tag{4.152}
$$

$$
\lambda_1 \left.\frac{\partial T_1}{\partial x_1}\right|_{x_1 = -0} = \lambda_2 \left.\frac{\partial T_2}{\partial x_1}\right|_{x_1 = +0}, \quad T_1|_{x_1 = -0} = T_2|_{x_1 = +0},
\tag{4.153}
$$

$$
\begin{aligned}
&T_1(0,x) = T_{1*}, \quad x_1 \leq 0, \quad u_1 \leq x_2 \leq u_2, \quad v_1 \leq x_3 \leq v_2, \\
&T_1(0,x) = T_{1,H}, \quad x_1 \leq 0, \quad 0 < x_2 < u_1, \quad u_2 < x_2 < L_2, \\
&\qquad 0 < x_3 < v_1, \quad v_2 < x_3 < L_3, \\
&T_2(0,x) = T_{2,H}, \quad x_1 > 0, \quad 0 < x_j < L_j,\ j = 2, 3,
\end{aligned}
\tag{4.154}
$$

$$
\begin{aligned}
T_1|_{x_1\to-\infty} &= T_{1*}, \quad u_1 \le x_2 \le u_2, \quad v_1 \le x_3 \le v_2,\\
T_1|_{x_1\to-\infty} &= T_{1,H}, \quad 0 < x_2 < u_1, \quad u_2 < x_2 < L_2,\\
&\quad 0 < x_3 < v_1, \quad v_2 < x_3 < L_3,\\
T_2|_{x_1\to\infty} &= T_{2,H};
\end{aligned}
\tag{4.155}
$$

$$
\frac{\partial T_i^{(2)}}{\partial t} = \xi_i \frac{\partial^2 T_i^{(2)}}{\partial x_2^2}, \quad i = 1, 2, \ 0 < t < t_*,
\tag{4.156}
$$

$$
\begin{aligned}
&T_i^{(2)}(0, x) = T_i^{(1)}(t_*, x), \quad T_i^{(2)}|_{x_2=0} = T_{i,H},\\
&\quad i = 1 : -\infty < x_1 \le 0, \quad i = 2 : 0 < x_1 < \infty,\\
&T_i^{(2)}|_{x_2=L_2} = T_{i,H}, \quad i = 1, 2, \quad 0 < x_3 < L_3;
\end{aligned}
\tag{4.157}
$$

$$
\frac{\partial T_i^{(3)}}{\partial t} = \xi_i \frac{\partial^2 T_i^{(3)}}{\partial x_3^2}, \quad i = 1, 2, \ 0 < t < t_*,
\tag{4.158}
$$

$$
\begin{aligned}
&T_i^{(3)}(0, x) = T_i^{(2)}(t_*, x), \quad T_i^{(3)}|_{x_3=0} = T_{i,H},\\
&\quad i = 1 : -\infty < x_1 \le 0, \quad i = 2 : 0 < x_1 < \infty,\\
&T_i^{(3)}|_{x_3=L_3} = T_{i,H}, \quad i = 1, 2, \ 0 < x_2 < L_2.
\end{aligned}
\tag{4.159}
$$

The analytical solution of a quasi-one-dimensional problem (4.152)–(4.155) is written in Section 4.1

$$
\begin{aligned}
&T_1^{(1)}(t_*, x) = T_{1*} + B_1\left[ERF\left(\frac{|x_1|}{2\sqrt{\xi_1 t_*}}\right) - 1\right], \quad -\infty < x_1 \le 0,\\
&B_1 = \frac{\sqrt{\xi_1}\lambda_2(T_{1*} - T_{2,H})}{\lambda_2\sqrt{\xi_1} + \lambda_1\sqrt{\xi_2}}, \quad u_1 \le x_2 \le u_2, \ v_1 \le x_3 \le v_2,\\
&T_1^{(1)}(t_*, x) = T_{1,H} + B_2\left[ERF\left(\frac{|x_1|}{2\sqrt{\xi_1 t_*}}\right) - 1\right], \quad -\infty < x_1 \le 0,\\
&B_2 = \frac{\sqrt{\xi_1}\lambda_2(T_{1,H} - T_{2,H})}{\lambda_2\sqrt{\xi_1} + \lambda_1\sqrt{\xi_2}}, \quad 0 < x_2 < u_1, \ u_2 < x_2 < L_2,\\
&0 < x_3 < v_1, \ v_2 < x_3 < L_3,
\end{aligned}
\tag{4.160}
$$

$$
\begin{aligned}
&T_2^{(1)}(t_*, x) = T_{1*} - B_1 + ERF\left(\frac{x_1}{2\sqrt{\xi_2 t_*}}\right)(B_1 - T_{1*} + T_{2,H}),\\
&0 < x_1 < \infty, u_1 \le x_2 \le u_2, v_1 \le x_3 \le v_2,
\end{aligned}
$$

$$T_2^{(1)}(t_*, x) = T_{1,H} - B_2 + ERF\left(\frac{x_1}{2\sqrt{\xi_2 t_*}}\right)(B_2 - T_{1,H} + T_{2,H}),$$
$$0 < x_1 < \infty,\ 0 < x_2 < u_1,\ u_2 < x_2 < L_2,$$
$$0 < x_3 < v_1,\ v_2 < x_3 < L_3. \tag{4.161}$$

The analytical solution of quasi-one-dimensional problems (4.156)–(4.159) looks like

$$T_i^{(2)}(x, t_*) = \left[T_i^{(1)}(t_*, x) + \int_0^{t_*} X_i^{(2)} \exp(\tau U_i^{(2)})\, d\tau\right]$$
$$\times \exp(-t_* U_i^{(2)}),\quad X_i^{(2)} = K_i^{(2)}/Y_i^{(2)},$$
$$i = 1:\ -\infty < x_1 \le 0,$$
$$i = 2:\quad 0 < x_1 < \infty,\quad 0 < x_3 < L_3, \tag{4.162}$$

$$K_1^{(2)} = (T_1^{(2)}|_{x_2=L_2} - T_1^{(2)}|_{x_2=0})x_2/L_2 + T_1^{(2)}|_{x_2=0},\quad -\infty < x_1 \le 0,$$
$$K_2^{(2)} = (T_1^{(2)}|_{x_2=L_2} - T_1^{(2)}|_{x_2=0})x_2/L_2 + T_1^{(2)}|_{x_2=0},\quad 0 < x_1 < \infty,$$
$$Y_i^{(2)} = \xi_i^{-1}\int_0^{L_2} G(x_2, y)\, dy,\quad U_i^{(2)} = 1/Y_i^{(2)},\quad i = 1, 2,$$

$$G(x_2, y) = \begin{cases} y(L_2 - x_2)/L_2, & 0 \le y \le x_2, \\ x_2(L_2 - y)/L_2, & x_2 \le y \le L_2. \end{cases} \tag{4.163}$$

$$T_i^{(3)}(x, t_*) = \left[T_i^{(2)}(t_*, x) + \int_0^{t_*} X_i^{(3)} \exp(\tau U_i^{(3)})\, d\tau\right]$$
$$\times \exp(-t_* U_i^{(2)}),\quad X_i^{(3)} = K_i^{(3)}/Y_i^{(3)},\ i = 1:\ -\infty < x_1 \le 0,$$
$$i = 2:\ 0 < x_1 < \infty,\ 0 < x_2 < L_2, \tag{4.164}$$

$$K_1^{(3)} = (T_1^{(3)}|_{x_3=L_3} - T_1^{(3)}|_{x_3=0})x_3/L_3 + T_1^{(3)}|_{x_3=0},\quad -\infty < x_1 \le 0,$$
$$K_2^{(3)} = (T_1^{(3)}|_{x_3=L_3} - T_1^{(3)}|_{x_3=0})x_3/L_3 + T_1^{(3)}|_{x_3=0},\quad 0 < x_1 < \infty,$$
$$Y_i^{(3)} = \xi_i^{-1}\int_0^{L_2} G(x_3, y)\, dy,\quad U_i^{(3)} = 1/Y_i^{(3)},\quad i = 1, 2,$$

$$G(x_3, y) = \begin{cases} y(L_3 - x_3)/L_3, & 0 \le y \le x_3, \\ x_3(L_3 - y)/L_3, & x_3 \le y \le L_3. \end{cases} \tag{4.165}$$

As a result we have the final solution of an initial boundary problem (4.146)–(4.151)

$$T_i(x, t_*) = T_i^{(3)}(x, t_*), \quad i = 1, 2. \tag{4.166}$$

According to the locally one-dimensional scheme algorithm of splitting [12, 13], the total analytical solution (4.166) is received by an exception of intermediate solution (4.160)–(4.164) of quasi-one-dimensional problems (4.152)–(4.159). Such an approach for the solution of the equations in partial derivatives with constants is offered and proved in [13].

## Existence, Uniqueness, and Convergence

Without loss of generality, we will consider the solution of the simplified boundary problem (4.85)–(4.89) with zero boundary conditions of the first type.

We take a nonlinear case $m > 0$, $k > 1$ in the region $\overline{Q_i} = Q_i + \Gamma$, $\overline{Q_{t,i}} = \overline{Q_i} \times [0 < t \leq t_0]$ at $Y_i = -1, C_i = 1, A_i(T_i) = T_i^m, A_{H,i} = 1, N_i = 1, E_i = 0, i = 1, 2, M = 0$

$$\frac{\partial T_1}{\partial t} = \sum_{j=1}^{3} \frac{\partial}{\partial x_j}\left[A_1(T_1)\frac{\partial T_1}{\partial x_j}\right] + B_1(T_1)\frac{\partial T_1}{\partial x_1} + \sum_{j=2}^{3} U_j(T_1)\frac{\partial T_1}{\partial x_j} - T_1^k,$$

$$\frac{\partial T_2}{\partial t} = \sum_{j=1}^{3} \frac{\partial}{\partial x_j}\left[A_2(T_2)\frac{\partial T_2}{\partial x_j}\right] + B_2(T_2)\frac{\partial T_2}{\partial x_1} + \sum_{j=2}^{3} F_j(T_2)\frac{\partial T_2}{\partial x_j} - T_2^k,$$

$$0 < t < t_*,$$

$$T_i(0, x) = T_{H,i}(x), \quad T_i|_\Gamma = 0, \quad i = 1, 2,$$

$$A_1\frac{\partial T_1}{\partial x_1}\bigg|_{x_1=-0} = A_2\frac{\partial T_2}{\partial x_1}\bigg|_{x_1=+0}, \quad T_1|_{x_1=-0} = T_2|_{x_1=+0}. \tag{4.167}$$

After application of the Kirhgof's transformation (4.90) at $A_{H,i} = 1, i = 1, 2$, formulas (4.106), (4.114)–(4.117) and locally one-dimensional scheme splittings (4.107)–(4.113), we have the modified boundary problem

$$\frac{\partial^2 w_i^{(1)}}{\partial x_1^2} + w_i^{(1)}\Phi_i = h_i^{(1)}, \quad 0 < t < t_*, \ i = 1, 2, \ n = 0, 1, 2, \ldots,$$

$$\eta\frac{\partial w_1^{(1)}}{\partial x_1}\bigg|_{x_1=-0} + Ew_0^{(1)} = \frac{\partial w_2^{(1)}}{\partial x_1}\bigg|_{x_1=+0}, \quad w_0^{(1)} = w_1^{(1)}|_{x_1=0},$$

$$\xi w_1^{(1)}|_{x_1=-0} = w_2^{(1)}|_{x_1=+0}, \quad w_i^{(1)}|_{t=0} = w_{H,i}, \quad i=1,2,$$
$$w_1^{(1)}|_{x_1=-a_1} = 0, \quad w_2^{(1)}|_{x_1=a_2} = 0, \quad 0 < x_j < L_j,\ j=2,3, \tag{4.168}$$

$$\frac{\partial^2 w_1^{(2)}}{\partial x_2^2} + w_1^{(2)}\Psi_2 = h_1^{(2)}, \quad -a_1 < x_1 < 0,$$
$$\frac{\partial^2 w_2^{(2)}}{\partial x_2^2} + w_2^{(2)}\Theta_2 = h_2^{(2)}, \quad 0 < x_1 < a_2,\ 0 < t < t_*,$$
$$w_i^{(2)}(0,x) = w_i^{(1)}(t_*,x), \quad w_i^{(2)}|_{x_2=0} = 0,$$
$$w_i^{(2)}|_{x_2=L_2} = 0, \quad i=1,2,\ -a_1 \le x_1 < 0,$$
$$0 \le x_1 \le a_2,\ 0 < x_3 < L_3, \tag{4.169}$$

$$\frac{\partial^2 w_1^{(3)}}{\partial x_3^2} + w_1^{(3)}\Psi_3 = h_1^{(3)}, \quad -a_1 < x_1 < 0,$$
$$\frac{\partial^2 w_2^{(3)}}{\partial x_3^2} + w_2^{(3)}\Theta_3 = h_2^{(3)}, \quad 0 < x_1 < a_2,\ 0 < t < t_*,$$

$$w_i^{(3)}(0,x) = w_i^{(2)}(t_*,x), \quad w_i^{(3)}|_{x_3=0} = 0,$$
$$w_i^{(3)}|_{x_3=L_3} = 0, \quad i=1,2,\ -a_1 \le x_1 < 0,$$
$$0 \le x_1 \le a_2,\ 0 < x_2 < L_2, \tag{4.170}$$

where $w_{H,i}, i = 1,2$ are taken from the last two formulas of relationships (4.108).

As a result of the application of the above described algorithm (4.120)–(4.145) to the boundary problem (4.168)–(4.170) we have solutions of the type (4.135), (4.136), (4.138), (4.140), (4.142), and (4.144), where in formulas (4.128), (4.132), (4.139), (4.141), (4.143), and (4.145) it is necessary to put $s_i = n_i = q_i = p_i = d_i = 0, i = 1,2$.

**Theorem.** *Let $w_i, i = 0,1,2$ be continuously differentiated in $\overline{Q_{t,i}}, i = 1,2$, then in the region $\overline{Q_{t,i}}, i = 1,2$ there is a unique solution of the modified boundary problem* (4.168)–(4.170).

Existence and uniqueness of the total solution (4.142)–(4.145) for a boundary problem (4.168)–(4.170) are proved in the same manner as was made in Chapter 2 [see formulas (2.19)–(2.71)].

## Estimation of Speed of Convergence [15, 16]

It is considered that in some neighborhood of a root function $f_i = f_i(w_i^n, \dot{w}_i^n), i = 1,2$ from Eq. (4.115) together with the partial derivatives $\partial f_i/\partial w_i,\ \partial^2 f_i/\partial w_i^2, \partial f_i/\partial \dot{w}_i, \partial^2 f_i/\partial \dot{w}_i^2, i = 1,2$ is continuous and $\partial f_i/\partial w_i$, $\partial^2 f_i/\partial w_i^2,\ \partial f_i/\partial \dot{w}_i,\ \partial^2 f_i/\partial \dot{w}_i^2, i = 1,2$ in this neighborhood does not go to zero.

Let's address the recurrence relationship (4.115) and noticing that $f_i(w_i, \dot{w}_i) = s_i(w_i) + r_i(\dot{w}_i), i = 1,2$ in Eq. (4.115), we will subtract $n$-e the equation from $(n+1)$th then we will find:

$$\frac{\partial^2(w_i^{(1)} - w_i^n)}{\partial x_1^2} = s_i(w_i^n) - s_i(w_i^{n-1}) - (w_i^n - w_i^{n-1})\frac{\partial s_i(w_i^{n-1})}{\partial w_i^{n-1}}$$

$$+ (w_i^{(1)} - w_i^n)\frac{\partial s_i(w_i^n)}{\partial w_i^n} + \left[ r_i(\dot{w}_i^n) - r_i(\dot{w}_i^{n-1}) - (\dot{w}_i^n \right.$$

$$\left. -\dot{w}_i^{n-1})\frac{\partial r_i(\dot{w}_i^{n-1})}{\partial \dot{w}_i^{n-1}} + (\dot{w}_i^{(1)} - \dot{w}_i^n)\frac{\partial r_i(\dot{w}_i^n)}{\partial \dot{w}_i^n} \right], \quad i = 1,2. \tag{4.171}$$

From the average theorem [33] it follows:

$$s_i(w_i^n) - s_i(w_i^{n-1}) - (w_i^n - w_i^{n-1})\frac{\partial s_i(w_i^{n-1})}{\partial w_i^{n-1}} = 0.5(w_i^n - w_i^{n-1})^2$$

$$\times \frac{\partial^2 s_i(\xi_i)}{\partial w_i^2}, \quad w_i^{n-1} \le \xi_i \le w_i^n, \quad i = 1,2.$$

Let's consider Eq. (4.171) and how the equation is in relation to $u_i^{(1)} = w_i^{(1)} - w_i^{(0)}, i = 1,2, (u_i^{(0)} = u_i^n, u_i^n = w_i^n - w_i^{n-1}, i = 1,2)$ and transform it as above Eqs. (4.118)–(4.145). Then we will have:

$$\frac{\partial^2 u_i^{(1)}}{\partial x_1^2} + u_i^{(1)}\Phi_i = h_i^{(1)}, \quad h_i^{(1)} = \dot{u}_i^{(1)}\frac{\partial r_i(\dot{w}_i^{(0)})}{\partial \dot{w}_i^{(0)}} + F_i^{(1)},$$

$$F_i^{(1)} = -0.5\sigma\left[ (u_i^{(0)})^2\frac{\partial^2 s_i(w_i^{(0)})}{\partial w_i^2} + (\dot{u}_i^{(0)})^2\frac{\partial^2 r_i(\dot{w}_i^{(0)})}{\partial \dot{w}_i^2} \right],$$

$$\Phi_i = -\frac{\partial s_i(w_i^{(0)})}{\partial w_i^{(0)}}, \quad u_i|_\Gamma = 0, \quad u_i(0,x) = 0, \quad i = 1,2, \quad \sigma = \frac{1}{3},$$

$$\eta \frac{\partial u_1^{(1)}}{\partial x_1}\bigg|_{x_1=-0} + Eu_0^{(1)} = \frac{\partial u_2^{(1)}}{\partial x_1}\bigg|_{x_1=+0}, \quad \xi u_1^{(1)}|_{x_1=-0} = u_2^{(1)}|_{x_1=+0},$$

$$u_i^{(1)} = \int_0^{a_i} G_i^{(1)}(x_1, y) h_i^{(1)}\, dy, \quad K_i^{(1)}(x,t) = 0, \quad V_1^{(1)}(a_1)$$
$$= V_2^{(1)}(a_2) = 0, \quad \dot{u}_i^{(1)} - u_i^{(1)} U_i^{(1)} = F_i^{(1)}/c_i, \quad U_i^{(1)} = 1/Y_i^{(1)},$$
$$Y_i^{(1)} = \int_0^{a_i} G_i^{(1)}(x_1, y) \frac{\partial r_i(\dot{w}_i^{(0)})}{\partial \dot{w}_i^{(0)}}\, dy, \quad c_i = \frac{\partial r_i(\dot{w}_i^{(0)})}{\partial \dot{w}_i^{(0)}}, \quad i = 1, 2, \quad (4.172)$$

where $G_i^{(1)}(x_1, y), i = 1, 2$ represent the formulas (4.130) at $i = 1\colon 0 \le x_1 \le a_1$ and (4.133) at $i = 2\colon\ 0 < x_1 \le a_2$.

The intermediate solution (4.172) to this problem in the first coordinate direction $x_1$ is written as

$$u_i^{(1)}(x, t_*) = \left[ u_{H,i} + \int_0^{t_*} X_i^{(1)}(w_i^{(0)}, x, \tau) \exp(-\tau U_i^{(1)})\, d\tau \right] \exp(t_* U_i^{(1)}),$$
$$u_i^{(0)} = u_i^n, \quad w_i^{(0)} = w_i^n, \ n = 0, 1, 2, \ldots, \ i = 1, 2, \ 0 < x_j < L_j, \ j = 2, 3,$$

$$X_i^{(1)} = -0.5\sigma \int_0^{a_i} G_i^{(1)}(x_1, y) \left[ (u_i^{(0)})^2 \frac{\partial^2 s_i(w_i^{(0)})}{\partial w_i^2} \right.$$
$$\left. + (\dot{u}_i^{(0)})^2 \frac{\partial^2 r_i(\dot{w}_i^{(0)})}{\partial \dot{w}_i^2}\, dy \right] / Y_i^{(1)}, \quad u_{H,i} = 0, \ i = 1, 2. \qquad (4.173)$$

The same possible solution (4.140) and (4.144) can be received on the co-ordinate directions $x_2$ and $x_3$ similar to Eq. (4.173). The final total solution will be:

$$u_i^{(3)}(x, t_*) = \exp(t_* U_i) \int_0^{t_*} X_i^{(1)} \exp(-\tau U_i^{(1)})\, d\tau + \exp[t_*(U_i^{(2)}$$
$$+ U_i^{(3)})] \int_0^{t_*} X_i^{(2)} \exp(-\tau U_i^{(2)})\, d\tau + \exp(t_* U_i^{(3)})$$
$$\times \int_0^{t_*} X_i^{(3)} \exp(-\tau U_i^{(3)})\, d\tau, \quad n = 0, 1, 2, \ldots, \ i = 1, 2,$$
$$(4.174)$$

$$U_i = \sum_{j=1}^{3} U_i^{(j)}(w^{(j-1)}), \quad U_i^{(j)}(w_i^{(j-1)}) = 1/Y_i^{(j)},$$

$$X_i^{(j)} = \frac{-0.5\sigma}{Y_i^{(j)}} \int_0^{L_j} G_i^{(j)}(x_j, y) \left[ (u_i^{(j-1)})^2 \frac{\partial^2 s_i(w_i^{(j-1)})}{\partial w_i^2} \right.$$

$$\left. +(\dot{u}_i^{(j-1)})^2 \frac{\partial^2 r_i(\dot{w}_i^{(j-1)})}{\partial \dot{w}_i^2} \right] dy, \quad u_i^{(0)} = u_i^{(n)}, \; i = 1,2,$$

$$Y_i^{(j)} = \int_0^{L_j} G_i^{(j)}(x_j, y) \frac{\partial r_i(\dot{w}_i^{(j-1)})}{\partial \dot{w}_i^{(j-1)}} dy, \quad j = 2,3,$$

where $G_i^{(j)}$ look like formulas (4.139) and (4.141) at $j = 2$, $i = 1,2$ and formulas (4.143) and (4.145) at $j = 3$, $i = 1,2$.

Let's put $\max\limits_{w_i, \dot{w}_i \in \overline{Q_{t,i}}} \left( \left| \frac{\partial^2 s_i(w_i^{(j)})}{\partial w_i^2} \right|, \left| \frac{\partial^2 r_i(\dot{w}_i^{(j)})}{\partial \dot{w}_i^2} \right| \right) = c_2$, $j = 0,1,2$, $\max\limits_{w_i, \dot{w}_i \in \overline{Q_{t,i}}} \left( \left| \frac{\partial s_i(w_i^{(j)})}{\partial w_i^{(j)}} \right|, \left| \frac{\partial r_i(\dot{w}_i^{(j)})}{\partial \dot{w}_i^{(j)}} \right| \right) = c_1$, $j = 0,1,2$, $\max\limits_{w_i \in \overline{Q_{t,i}}} (|\Phi_i|, |\Psi_j|, |\Theta_j|) = c_3, j = 2,3, i = 1,2$, $\max\limits_{x_1, y} |G_i^{(1)}(x_1, y)| \le |b_i^{-1} \sin(b_1 y)| \le b_i^{-1}$, $b_i^2 = \Phi_i$, $i = 1,2$, $\max\limits_{x_j, y} |G_1^{(j)}(x_j, y)| \le \beta_j^{-1}, \beta_j^2 = \Psi_j$, $\max\limits_{x_j, y} |G_2^{(j)}(x_j, y)| \le \gamma_j^{-1}, \gamma_j^2 = \Theta_j, j = 2,3$, $\eta_i = \max\limits_i (a_i, L_2, L_3), i = 1,2$, assuming that $c_m < \infty$, $m = 1,2,3$. Then noting that $\partial^2 r_i(\dot{w}_i^{(j)})/\partial \dot{w}_i^2 = 0, i = 1,2, j = 0,1,2$ and using the assumption about equivalence of all directions in space ($U_i^{(j)} = \alpha_i, j = 1,2,3, X_i^{(j)} = B(u_i^n)^2, i = 1,2, j = 1,2,3$) and equivalent functions $u_i^{(0)} = u_i^{(j)}, j = 1,2, i = 1,2$ (for converging sequences $w_i^n, i = 1,2$ all intermediate values $u_i^{(j)}, i = 1,2, j = 0,1,2$ are close to zero as they are in a convergence interval: $[w_i^{(0)}, w_i^{(3)}], i = 1,2$), we have from Eq. (4.174) at $u_i^{n+1}(t_*, x) = u_i^{(3)}(t_*, x), i = 1,2$, omitting an index $(*)$ at $t$ below and introducing $\alpha_i = \sqrt{c_3}/(c_1\eta_i), b_i = \beta_j = \gamma_j = \sqrt{c_3}, i = 1,2, j = 2,3$, $B = 0.5\sigma c_2/c_1$

$$|u_i^{n+1}| \le BV_i(u_i^n)^2 \int_0^t \exp(-\alpha_i \tau)\, d\tau, \quad i = 1,2,$$

$$-a_1 < x_1 < a_2, \; 0 < x_j < L_j, \; j = 2,3,$$

$$V_i = \exp(3\alpha_i t) + \exp(2\alpha_i t) + \exp(t\alpha_i), \quad i = 1,2,$$

$$|u_i^{n+1}| \le BV_i(u_i^n)^2[1 - \exp(-\alpha_i t)]/\alpha_i, \quad i = 1,2. \tag{4.175}$$

Let's choose $u_i^0(t,x)$ so that $|u_i^0(t,x)| \leq 1, i = 1,2$ in area $\overline{Q_{t,i}}, i = 1,2$. As a result from expression (4.175) we will receive at $n = 0$, introducing $M_i^1 = \max\limits_{\overline{Q_{t,i}}}|u_i^1|$, $z_i^0 = \max\limits_{\overline{Q_{t,i}}}(u_i^0)^2, z_i^0 \leq 1$

$$M_i^1 \leq \frac{B[\exp(3t\alpha_i) - 1]}{\alpha_i} = S_i, \quad i = 1,2. \tag{4.176}$$

Hence, at $\alpha_i > 0$ we see that the top border $M_i^1, i = 1,2$ will not surpass 1 if there is inequality $S_i \leq 1, i = 1,2$ in Eq. (4.176):

$$t \leq \ln\left(\frac{\alpha_i}{B} + 1\right)^{1/3\alpha_i}, \quad i = 1,2. \tag{4.177}$$

Therefore choosing an interval $[0,t]$ small enough so that the condition is satisfied from Eq. (4.177), we will have $M_i^1 \leq 1, i = 1,2$. Finally we receive $M_i^{n+1} \leq S_i z_i^n, i = 1,2$ or

$$\max_{x,t\in\overline{Q_{t,1}}} |w_1^{n+1} - w_1^n| \leq S_1 \max_{x,t\in\overline{Q_{t,1}}} |w_1^n - w_1^{n-1}|^2, \tag{4.178}$$

$$\max_{t\in\overline{Q_{t,2}}} |w_2^{n+1} - w_2^n| \leq S_2 \max_{t\in\overline{Q_{t,2}}} |w_2^n - w_2^{n-1}|^2. \tag{4.179}$$

Relationships (4.178) and (4.179) show that if convergence of the iterative process for a boundary problem (4.168)–(4.170) or (4.167) according to the inversion formula (4.114) takes place, it is quadratic. Thus, at a big enough $n$ each following step doubles a number of correct signs in the given approximation.

## Results of Test Checks

Estimation of an error of analytical formulas (4.114), (4.127)–(4.136), (4.138)–(4.145) are checked practically in the solution of a boundary problem (4.85)–(4.89):

$$C_1(T_1)\frac{\partial T_1}{\partial t} = \sum_{j=1}^{3} \frac{\partial}{\partial x_j}\left[A_1(T_1)\frac{\partial T_1}{\partial x_j}\right] + B_1(T_1)\frac{\partial T_1}{\partial x_1}$$
$$+ \sum_{j=2}^{3} U_j(T_1)\frac{\partial T_1}{\partial x_j} + Y_1 T_1^k + E_1(x,t),$$

$$C_2(T_2)\frac{\partial T_2}{\partial t} = \sum_{j=1}^{3}\frac{\partial}{\partial x_j}\left[A_2(T_2)\frac{\partial T_2}{\partial x_j}\right] + B_2(T_2)\frac{\partial T_2}{\partial x_1}$$
$$+\sum_{j=2}^{3}F_j(T_2)\frac{\partial T_2}{\partial x_j} + Y_2T_2^k + E_2(x,t) \qquad (4.180)$$

in the same range of definition $\overline{Q_{t,i}}$, $i = 1, 2$ with the initial condition

$$T_i\,|_{t=0} = \exp(y_1 + y_2 + y_3),\ i = 1,2, \quad y_1 = |x_1/a_1|, \quad -a_1 < x_1 < 0,$$
$$y_1 = x_1/a_2, \quad 0 < x_1 < a_2, \quad y_j = x_j/L_j, \quad j = 2,3, \qquad (4.181)$$

with the matching condition

$$A_1\frac{\partial T_1}{\partial x_1}\bigg|_{x_1=-0} = A_2\frac{\partial T_2}{\partial x_1}\bigg|_{x_1=+0}, \qquad T_1|_{x_1=-0} = T_2|_{x_1=+0} \qquad (4.182)$$

and with boundary conditions of the first type

$$T_1|_{x_1=-a_1} = \exp(\tau + 1 + y_2 + y_3), \quad T_2|_{x_1=a_2} = \exp(\tau + 1 + y_2 + y_3),$$
$$T_i|_{x_2=0} = \exp(\tau + y_1 + y_3), \quad T_i|_{x_2=L_2} = \exp(\tau + y_1 + 1 + y_3),$$
$$T_i|_{x_3=0} = \exp(\tau + y_1 + y_2), \quad T_i|_{x_3=L_3} = \exp(\tau + y_1 + y_2 + 1),$$
$$\tau = t/c, \quad i = 1,2. \qquad (4.183)$$

The exact solution of a problem (4.180)–(4.183) was taken

$$T_i = \exp(\tau + z), \quad z = \sum_{j=1}^{3} y_j, \quad i = 1,2, \qquad (4.184)$$

then sources $E_i, i = 1,2$ in Eq. (4.180) look like at $A_i = N_iT_i^m, i = 1,2$

$$E_1 = \exp(\tau + z)\left\{\frac{C_1}{c} - \left(\frac{B_1}{a_1} + \frac{U_2}{L_2} + \frac{U_3}{L_3}\right)\right.$$
$$\left.-s\left(\frac{1}{a_1^2} + \frac{1}{L_2^2} + \frac{1}{L_3^2}\right)N_1\exp[m(\tau + z)]\right\}$$
$$-\ Y_1\exp[k(\tau + z)], \quad s = m + 1,$$

$$E_2 = \exp(\tau + z)\left\{\frac{C_2}{c} - \left(\frac{B_2}{a_2} + \frac{F_2}{L_2} + \frac{F_3}{L_3}\right)\right.$$
$$\left.-s\left(\frac{1}{a_2^2} + \frac{1}{L_2^2} + \frac{1}{L_3^2}\right)N_2\exp[m(\tau + z)]\right\}$$
$$-\ Y_2\exp[k(\tau + z)].$$

The following basic values of initial data were used: $m = 0.5$, $t_0 = 1$, $A_{H,i} = 1, N_i = a_i, c = 10, \Delta x_1 = a_i/(N-1), i = 1,2, \Delta x_j = L_j/(M-1), j = 2,3, (N = M = 11), \Delta t = 0.02$ is a number of checkouts and steps on space and time while finding integrals in the Eqs. (4.128)–(4.136), (4.138)–(4.145) by Simpson's formula [21].

The boundary problem (4.180)–(4.183) is solved by means of formulas (4.114), (4.127)–(4.136), (4.138)–(4.145). The number of iterations was traced [for total expressions of a type (4.142)–(4.145)] on relative change of an error vector:

$$||V_i^n|| = \max_{xi,t\in\overline{Q_{t,i}}} \left| \frac{w_i^{n+1} - w_i^n}{w_i^{n+1}} \right|, \quad i = 1,2.$$

The program is made in the language Fortran-90, calculation was made on Pentium 5 (3.5 GHz, compiler PS 5) with double accuracy. In Table 4.2 maximum relative error is given: $\varepsilon = \max(\varepsilon_1, \varepsilon_2)$

$$\varepsilon_i = \frac{|T_i - \tilde{T}_i|100\%}{T_i}, \quad i = 1,2, \tag{4.185}$$

where $T_i$ is the exact obvious solution (4.184), $\tilde{T}_i, i = 1,2$ is the approximate analytical solution on mathematical technology of this paragraph at various values $k$, $B_i$, $U_i$, $Y_i$, $i = 1,2$, $U_j = F_j, j = 2,3$. In Eq. (4.185) errors $\varepsilon_i, i = 1,2$ answer $a_1 = 0.6$ m, $a_2 = 0.55$ m, $L_2 = L_3 = 0.07$ m, $C_1 = 3.4 \cdot 10^3$ s/m$^2$ $C_2 = 8.61 \cdot 10^3$ s/m$^2$ (a material of type of a graphite V–1 and copper [31]).

In Table 4.2 the results of test calculations $||V_i^n|| \le \delta, i = 1,2$ are given $\delta = 0.01$. Thus, only two iterations were required to achieve this accuracy, and the time of calculation of any variant was $t_p = 2$ s.

As we can see from Table 4.2 calculation on developed MT has almost the small error $\varepsilon = \max(\varepsilon_1, \varepsilon_2)$ from Eq. (4.185).

In comparison with the numerical solution of a problem (4.180)–(4.183) locally-one-dimensional scheme splitting [12, 13] of the type (4.99)–(4.105) was used. For the solution of the quasi-one-dimensional equations of this system we used the scheme of implicit absolutely steady differential with approximation error (in total sense [12]) for the first and second derivative on space—$O\left[\sum_{j=1}^{3}(\Delta x_j)^2\right]$ and the two-layer scheme for derivative on time with a margin error approximations—$O(\Delta t)$.

For differential grids $\Delta x_1 = a_i/(N-1), i = 1,2, \Delta x_j = L_j/(M-1), j = 2,3, (N = M = 21), \Delta t = 0.02$ and other identical initial data for a variant

**Table 4.2** A dependence of the maximum relative errors at the various values $k, B_i, Y_i, i = 1, 2, U_i, i = 2, 3$

| | Results of calculations | | | | | | | |
|---|---|---|---|---|---|---|---|---|
| **Variant number** | $k$ | $B_1$ | $B_2$ | $U_2$ | $U_3$ | $Y_1$ | $Y_2$ | $\varepsilon, \%$ |
| 1 | 1 | −1 | −1 | −1 | −1 | −1 | −1 | 5.2 |
| 2 | 2 | −1 | −1 | −1 | −1 | −1 | −1 | 5.15 |
| 3 | 1 | 1 | 1 | −1 | −1 | −1 | −1 | 9.98 |
| 4 | 2 | 1 | 1 | −1 | −1 | −1 | −1 | 9.99 |
| 5 | 1 | 1 | 1 | 1 | 1 | −1 | −1 | 9.22 |
| 6 | 1 | −1 | −1 | −1 | −1 | −100 | −100 | 4.55 |
| 7 | 1 | −1 | −1 | −1 | −1 | 30 | 30 | 5.42 |

at number 1 of Table 4.2 the numerical solution gives, during the moment, $t_0 = 1$ s $\varepsilon = 6.48\%$ and $t_p = 4$ s.

As we can see from Table 4.2, accuracy of calculation on developed mathematical technology of this paragraph is not worse than numerical.

Let's consider now the received formulas with reference to the solution of a problem of heat-exchange (4.146)–(4.151) at $a_1 = 0.95$ m, $a_2 = 0.9$ m, $L_2 = L_3 = 0.5$ m, $\xi_1 = 1.16 \cdot 10^{-4}\,\text{m}^2/\text{s}$, $A_1 = \lambda_1 = 386\,\text{W}/(\text{m} \cdot \text{K})$, $\xi_2 = 5 \cdot 10^{-6}\,\text{m}^2/\text{s}$, $A_2 = \lambda_2 = 23\,\text{W}/(\text{m} \cdot \text{K})$.

Let's estimate analytical formulas (4.51)–(4.56) from the first coordinate direction $x_1$ from Section 4.1, Eqs. (4.162)–(4.166) from the second and third direction $x_2, x_3$ in the problem solution (4.146)–(4.151) and compare their accuracy with the analytical solution of this problem on the Eqs. (4.160)–(4.166). It is obvious in comparison that in Eq. (4.185) total expressions (4.164)–(4.166), which are received as a result of an exception of the intermediate solutions from the previous coordinate directions, $x_1, x_2$ are substituted. The difference of temperature on joints of materials 1 and 2 in a plane $x_1 = 0$ at the moment of time $t_0 = 81$ s is not more $\varepsilon = 4.83\%$.

# CHAPTER 5

# Method of Solution of Equations in Partial Derivatives

In mathematical modeling, high-intensity processes of heat propagation [47, 48], processes of electromagnetic oscillations [18], etc. give rise to problem solving in the partial derivatives of the equations of the second order hyperbolic type. Here, over a time of thermal inertia the finite speed of propagation of heat in the boundary conditions of the third kind is counted. In a chapter on generalization algorithms [16] the approximate analytical solutions for the nonlinear one-dimensional and three-dimensional telegraph equation were discussed. Comparison of accuracy of the analytical formula of the one-dimensional telegraphic equation developed by mathematical technology with known analytical solution of this problem is given in [49]. In the three-dimensional case, by the trial function test checks of mathematical technology, results were achieved and a comparison with the known numerical method is given.

## 5.1 METHOD OF SOLUTION OF ONE-DIMENSIONAL THERMAL CONDUCTIVITY HYPERBOLIC EQUATION

### Statement of a Problem and a Method Algorithm

Let us attempt to find a solution of the energy equation [47], where the heat flux is given as [50]; in the case of one-dimensional we have:

$$C(T)\frac{dT}{dt} = -\frac{\partial q}{\partial x} + F(t,x), \quad 0 < t \le t_0, \tag{5.1}$$

$$q = -A(T)\frac{\partial T}{\partial x} - \tau_r \frac{\partial q}{\partial t}. \tag{5.2}$$

From Eq. (5.1) we obtain

$$\frac{\partial q}{\partial x} = F(t,x) - C\frac{dT}{dt}. \tag{5.3}$$

*Analytical Solution Methods for Boundary Value Problems*
http://dx.doi.org/10.1016/B978-0-12-804289-2.00005-3

We substitute Eq. (5.2) in Eq. (5.1), then

$$C\frac{dT}{dt} = \frac{\partial}{\partial x}\left[A(T)\frac{\partial T}{\partial x}\right] + \tau_r\frac{\partial}{\partial t}\frac{\partial q}{\partial x} + F(t,x). \tag{5.4}$$

Assuming that $C$ for the final (fast) time interval $t \approx \tau_r$ ($\tau_r$—relaxation time) varies slightly ($C \approx$ const) we obtain from Eqs. (5.3) and (5.4), painting the total derivative $dT/dt$

$$C\left(\frac{\partial T}{\partial t} + w\frac{\partial T}{\partial x}\right) + \tau_r C\left(\frac{\partial^2 T}{\partial t^2} + w\frac{\partial^2 T}{\partial t \partial x}\right)$$
$$= \frac{\partial}{\partial x}\left[A(T)\frac{\partial T}{\partial x}\right] + \tau_r\frac{\partial F}{\partial t} + F(x,t). \tag{5.5}$$

Following [47], consider that $C(T), w(T)$ is clearly not dependent on time, space and $w < 1\,\text{m/s}$ neglect mixed derivative $w\frac{\partial^2 T}{\partial t \partial x}$ in comparison with $\frac{\partial^2 T}{\partial t^2}$ in the left-hand side of Eq. (5.5). Furthermore, in Eq. (5.5) to the right-side of the term is not considered $\tau_r\frac{\partial F}{\partial t}$ [$\tau_r = \chi/c^2$, $\chi$—the thermal diffusivity, $\text{m}^2/\text{s}$, $c =$ const, $c$—the velocity of propagation thermal perturbation (the speed of sound in the medium), m/s], the value of which for time discussed below: $t > 100 \cdot \tau_r$ ($\tau_r \sim 10^{-9}$ s) is negligible.

If the source is present in an additional nonlinear term, the result is in solutions of hyperbolic equations of second order [18, 47]

$$C(T)\frac{\partial T}{\partial t} + z^{-1}\frac{\partial^2 T}{\partial t^2} = \frac{\partial}{\partial x}\left[A(T)\frac{\partial T}{\partial x}\right] + Y(T)\frac{\partial T}{\partial x} + A_1 T^k + A_2(x,t) \tag{5.6}$$

in $Q = (0 < x < a, 0 < a < \infty), \overline{Q} = Q + \Gamma, \overline{Q}_t = \overline{Q} \times [0 < t \le t_0]$, $\Gamma$—boundary surface domain of the $Q$ with the initial conditions

$$T|_{t=0} = p_1(x), \qquad \left.\frac{\partial T}{\partial t}\right|_{t=0} = p_2(x), \tag{5.7}$$

where $A_1 =$ const, $Y(T) = -C(T)w(T)$, $z = c^2/A_H$, $A_H =$ const, $w$—the rate of convective heat transfer, m/s; $C(T)$—the coefficient of volumetric heat capacity, $\text{J}/(\text{K} \cdot \text{m}^3)$; $A(T)$—thermal conductivity, $A_H$—factor thermal conductivity at the initial temperature, $\text{W}/(\text{K} \cdot \text{m})$; $T$—temperature, K; $a$—length of the segment in space, m; $t_0$—end of the range, s.

A.V. Lykov [47], analyzing the generalized problem of heat conduction for a half-space, boundary the temperature which is changed at the initial time by certain amount, then remaining constant, justifies the physical meaning of the finite speed propagation of heat as the time derivative of the depth of penetration of heat.

For Eq. (5.6) are more commonly used classical boundary conditions (3.3) (see Chapter 3, Section 3.1), and not integrated (or differential) form of writing these conditions arising from the Eq. (5.2). On the feasibility of compliance with the boundary conditions for the Eq. (5.6) law (5.2) previously ignored authors [51, 52]. Then, following [52], we obtain

$$q|_{\Gamma} = -A\frac{\partial T}{\partial x}\bigg|_{\Gamma} - \tau_r\frac{\partial q}{\partial t}\bigg|_{\Gamma}, \quad t > 0. \tag{5.8}$$

Taking into account that the heat flux through the boundary $\Gamma$ field Q, in accordance with Newton's law [37], is proportional to the temperature difference between the border and the external environment, we obtain ($\alpha$—heat transfer coefficient, $W/(K \cdot m^2)$)

$$q|_{\Gamma} = \alpha(T - D)|_{\Gamma}, \quad t > 0. \tag{5.9}$$

We express from Eq. (5.8) the value of $q|_{\Gamma}$ in terms of temperature, given that the initial time point $q(0, \Gamma) = 0$. Then, according to [26] we have:

$$q(t, \Gamma) = \left[B - \frac{A}{\tau_r}\int_0^t \frac{\partial T}{\partial x}\bigg|_{\Gamma} \exp\left(\frac{\tau}{\tau_r}\right) d\tau\right] \exp\left(-\frac{t}{\tau_r}\right),$$
$$B = q(0, \Gamma), \quad t > 0 \tag{5.10}$$

or in differential form ratio (5.10) can be rewritten according to [52], using the formula (5.9)

$$A\frac{\partial T}{\partial x}\bigg|_{\Gamma} = -\alpha S_t(T - D)|_{\Gamma}, \quad S_t = 1 + \tau_r\frac{\partial}{\partial t}, \quad t > 0. \tag{5.11}$$

At $\alpha = 0$ we receive Neumann's condition, and at $\alpha \to \infty$ we receive Dirichlet's condition [boundary condition of the 1st type (5.68)].

Let's assume everywhere:

1. A problem (5.6), (5.7), and (5.11) has a unique solution $T(x, t)$, which is continuously in the closed region $\overline{Q}_t$ and has continuous derivatives $\frac{\partial T}{\partial t}, \frac{\partial^2 T}{\partial t^2}, \frac{\partial T}{\partial x}, \frac{\partial^2 T}{\partial x^2}$.
2. The following conditions are satisfied: $A(T) \geq c_1 > 0$, $C(T) \geq c_2 > 0$, $z \geq c_3 > 0$, $c_1, c_2, c_3$ are constants; $T_H$ is continuous given function $\overline{Q}$, and $A, Y, A_2$ are the continuous functions in the closed area $\overline{Q}_t$.

3. Generally, coefficients $C(T), Y(T)$ can be nonlinearly dependent on the solution of the problem $T$ [5], form $A(T)$ is defined below in the formula (5.19), and $D$ is the given continuous function on $\Gamma$ for $0 < t \leq t_0$, having limited partial derivatives of the first order.

We introduce for the problem (5.6), (5.7), and (5.11) intermediate steps of the solution: the direction of (1) is the solution of the wave and the direction of (2) is the parabolic part of the solution of the original Eq. (5.6). Then we have:

$$z^{-1}\frac{\partial^2 T^{(1)}}{\partial t^2} = \xi\frac{\partial}{\partial x}\left[A(T^{(0)})\frac{\partial T^{(1)}}{\partial x}\right] + \xi A_2(x,t), \quad 0 < t < t_*, \tag{5.12}$$

$$T^{(0)}|_{t=0} = p_1(x), \quad (\partial T^{(0)}/\partial t)|_{t=0} = p_2(x), \tag{5.13}$$

$$\left[A\frac{\partial T^{(1)}}{\partial x} + \alpha_1 S_t T^{(1)}\right]_{x=0} = \alpha_1 S_t D_1(t),$$
$$\left[A\frac{\partial T^{(1)}}{\partial x} + \alpha_2 S_t T^{(1)}\right]_{x=a} = \alpha_2 S_t D_2(t,a); \tag{5.14}$$

$$C(T^{(1)})\frac{\partial T^{(2)}}{\partial t} = \eta\frac{\partial}{\partial x}\left[A(T^{(1)})\frac{\partial T^{(2)}}{\partial x}\right] + Y(T^{(1)})\frac{\partial T^{(2)}}{\partial x}$$
$$+ A_1(T^{(2)})^k + \eta A_2(x,t), \quad 0 < t < t_*, \tag{5.15}$$

$$T^{(2)}|_{t=0} = T^{(1)}(t_*, x), \tag{5.16}$$

$$\left[A\frac{\partial T^{(2)}}{\partial x} + \alpha_1 S_t T^{(2)}\right]_{x=0} = \alpha_1 S_t D_1(t),$$
$$\left[A\frac{\partial T^{(2)}}{\partial x} + \alpha_2 S_t T^{(2)}\right]_{x=a} = \alpha_2 S_t D_2(t,a), \tag{5.17}$$

where $\xi + \eta = 1$. If the $C(T) \neq 0$ telegraph equation is solved (5.12)–(5.17): friction loss (the conductive medium), if $\eta = 0, C(T) = 0, A_1 = 0$ is solved by wave equation (5.12)–(5.14): lack of friction (the decaying medium).

We will use the inversion formula, $A(T)$ in Eq. (5.12) taken in the form

$$A(T) = NT^m, \quad m > 0,\ N > 0,\ N = \text{const.} \tag{5.18}$$

Let's use Kirchhoff's transformation [5] to Eqs. (5.12)–(5.17)

$$v = \int_0^T \frac{A(T)}{A_H}\, dT, \tag{5.19}$$

where $A_H$ is, for example, the coefficient of thermal conductivity at a temperature equal to zero.

Then, taking into account the relationships [5]:

$$\frac{\partial A}{\partial x} = \frac{\partial A}{\partial T}\frac{\partial T}{\partial x}, \quad \frac{\partial v}{\partial t} = \frac{A}{A_H}\frac{\partial T}{\partial t}, \quad \frac{\partial v}{\partial x} = \frac{A}{A_H}\frac{\partial T}{\partial x}, \tag{5.20}$$

we receive a boundary problem from Eqs. (5.12)–(5.20):

$$b_\xi \frac{\partial}{\partial t}\left(\frac{A_H}{A}\frac{\partial v^{(1)}}{\partial t}\right) = \frac{\partial^2 v^{(1)}}{\partial x^2} + a_2(x,t), \quad 0 < t < t_*, \tag{5.21}$$

$$T^{(0)}|_{t=0} = p_1(x), \quad (\partial T^{(0)}/\partial t)|_{t=0} = p_2(x), \tag{5.22}$$

$$\left[\frac{\partial v^{(1)}}{\partial x} + \frac{\alpha_1}{A_H} S_t(\phi v^{(1)})^{1/s}\right]_{x=0} = \frac{\alpha_1}{A_H} S_t D_1,$$

$$\left[\frac{\partial v^{(1)}}{\partial x} + \frac{\alpha_2}{A_H} S_t(\phi v^{(1)})^{1/s}\right]_{x=a} = \frac{\alpha_2}{A_H} S_t D_2, \tag{5.23}$$

$$\frac{\partial^2 v^{(2)}}{\partial x^2} = c_\eta \frac{\partial v^{(2)}}{\partial t} - \gamma_\eta \frac{\partial v^{(2)}}{\partial x} - a_1(\phi v^{(2)})^{k/s} - a_2(x,t), \quad 0 < t < t_*, \tag{5.24}$$

$$v^{(2)}(0,x) = v^{(1)}(t_*,x), \quad v^{(1)} = \frac{[T^{(1)}(t_*,x)]^s}{\phi},$$

$$\phi = sA_H/N, \quad s = m+1, \tag{5.25}$$

$$\left[\frac{\partial v^{(2)}}{\partial x} + \frac{\alpha_1}{A_H} S_t(\phi v^{(2)})^{1/s}\right]_{x=0} = \frac{\alpha_1}{A_H} S_t D_1,$$

$$\left[\frac{\partial v^{(2)}}{\partial x} + \frac{\alpha_2}{A_H} S_t(\phi v^{(2)})^{1/s}\right]_{x=a} = \frac{\alpha_2}{A_H} S_t D_2, \tag{5.26}$$

where $a_2 = A_2/A_H, b_\xi = 1/(\xi c^2)$, $a_1 = A_1/(\eta A_H), \gamma_\eta = Y(T^{(1)})/(\eta A)$, $c_\eta = C(T^{(1)})/(\eta A)$.

In this case $T^{(i)}, i = 1, 2$ is determined from Eq. (5.19) according to the inversion formula

$$v^{(i)} = (T^{(i)})^s/\phi, \quad T^{(i)} = (v^{(i)}\phi)^{1/s}, \quad i = 1, 2. \tag{5.27}$$

Let's note that variation ranges of independent variables and type of the boundary conditions do not change in relation to Kirchhoff's transformation (5.19), and within the inversion formula (5.27) the boundary conditions first-third types pass into Dirichlet, Neumann, and Newton's conditions.

Our purpose is to receive a solution of a nonlinear boundary problem, if it exists, as a limit of sequence of solutions of the linear boundary problems. For this we use the results of works [7, 15, 16]. Let $v^{(1)} = \text{const}$ be some initial approximation [as an initial approximation it is expedient to take $v^{(1)}(t_*, x)$ of Eq. (5.25)]. We will consider a sequence $v_n(t, x)$, defined by the recurrence relation [7] (the dot and bar at the top denotes the partial derivative with respect to time and space):

$$\frac{\partial^2 v_{n+1}}{\partial x^2} = f + (v_{n+1} - v_n)\frac{\partial f}{\partial v_n} + (\dot{v}_{n+1} - \dot{v}_n)\frac{\partial f}{\partial \dot{v}_n},$$
$$+ (v'_{n+1} - v'_n)\frac{\partial f}{\partial v'_n}, \quad f = f(v_n, \dot{v}_n, v'_n), \tag{5.28}$$

$$v^{(2)}(0, x) = v^{(1)}(t_*, x), \quad \left.\frac{\partial v^{(j)}}{\partial x}\right|_{\Gamma}$$
$$= \left[\beta + (v^{(j)} - v^{(j-1)})\frac{\partial \beta}{\partial v^{(j-1)}}\right]_{\Gamma}, \quad j = 1, 2,$$
$$Z_j = \phi v^{(j-1)}, \quad \beta = \frac{\alpha}{A_H} S_t (D - Z_j^{1/s})|_{\Gamma},$$

or

$$\left.\left(\frac{\partial v^{(j)}}{\partial x} + v^{(j)}\frac{\alpha}{N} S_t Z_j^{-m/s}\right)\right|_{\Gamma} = \left.\frac{\alpha}{A_H}\left(S_t D - \frac{m}{s} S_t Z_j^{1/s}\right)\right|_{\Gamma}, \quad j = 1, 2. \tag{5.29}$$

Each function $v_{n+1}(t, x)$ in Eqs. (5.28) and (5.29) is a solution of the linear equation that is a rather important feature of this algorithm. The algorithm comes from an approximation of Newton-Kantorovich's method [8] in functional space.

To reduce further records we will introduce the notations:

$$f = c_\eta \dot{v}^{(1)} - \gamma_\eta (v^{(1)})' - a_1 (Z_2)^{k/s} - a_2(t,x), \quad Z_j = \phi v^{(j-1)},$$

$$R_\xi = a_2(t,x), \quad \Phi_\eta = \frac{a_1 k (Z_2)^{k/s} A_H}{Z_2 N}, \quad \gamma_\eta = \frac{Y(T)}{\eta A(T)}, \quad j = 1, 2,$$

$$R_\eta = a_1 (Z_2)^{k/s} \left(1 - \frac{v^{(1)} k A_H}{Z_2 N}\right) + a_2, \quad X_j = \phi v^{(j-1)}(t,0),$$

$$Y_j = \phi v^{(j-1)}(t,a), \quad j = 1, 2,$$

$$c_\eta = \frac{\partial f}{\partial \dot{v}^{(1)}}, \quad \gamma_\eta = -\frac{\partial f}{\partial (v')^{(1)}}, \quad \Phi_\eta = -\frac{\partial f}{\partial v^{(1)}}, \quad h_2 = c_\eta \dot{v}^{(2)} - R_\eta (v^{(1)}),$$

$$g_1^{(j)} = \frac{\alpha_1}{N} X_j^{-m/s} \left(1 - \frac{\tau_r m \phi}{s X_j} \frac{\partial v^{(j-1)}}{\partial t}\right)\Bigg|_{x=0},$$

$$g_2^{(j)} = \frac{\alpha_2}{N} Y_j^{-m/s} \left(1 - \frac{\tau_r m \phi}{s Y_j} \frac{\partial v^{(j-1)}}{\partial t}\right)\Bigg|_{x=a}, \quad j = 1, 2,$$

$$q_1^{(j)} = \frac{\alpha_1}{A_H} \left[ S_t D_1 - \frac{m}{s} X_j^{1/s} \left(1 + \frac{\tau_r \phi}{s X_j} \frac{\partial v^{(j-1)}}{\partial t}\right)\right]\Bigg|_{x=0},$$

$$q_2^{(j)} = \frac{\alpha_2}{A_H} \left[ S_t D_2 - \frac{m}{s} Y_j^{1/s} \left(1 + \frac{\tau_r \phi}{s Y_j} \frac{\partial v^{(j-1)}}{\partial t}\right)\right]\Bigg|_{x=a}, \quad j = 1, 2, \quad (5.30)$$

where $g_i^{(1)}$ and $q_i^{(1)}$ : $i = 1$ at $x = 0$, $i = 2$ at $x = a$ is wave part of the telegraph equation (5.6) and $g_i^{(2)}$ and $q_i^{(2)}$ : $i = 1$ at $x = 0$, $i = 2$ for $x = a$ is parabolic part equation (5.6).

Let's take the solution of a boundary problem (5.28)–(5.29), using the Eqs. (5.21)–(5.23) and formulas (5.27) and (5.30), then we have

$$\frac{\partial^2 v^{(1)}}{\partial x^2} = h_1, \quad h_1 = b_\xi \frac{\partial}{\partial t} \left( \frac{A_H}{A} \frac{\partial v^{(1)}}{\partial t} \right) - a_2, \quad 0 < t < t_*, \quad (5.31)$$

$$T^{(0)}|_{t=0} = p_1(x), \quad (\partial T^{(0)}/\partial t)|_{t=0} = p_2(x), \quad (5.32)$$

$$\left[\frac{\partial v^{(1)}}{\partial x} + g_1^{(1)} v^{(1)}\right]_{x=0} = q_1^{(1)}, \quad \left[\frac{\partial v^{(1)}}{\partial x} + g_2^{(1)} v^{(1)}\right]_{x=a} = q_2^{(1)}. \quad (5.33)$$

It is supposed that in calculation of images on coordinate $x$, we operate with functions, analytically continued on values $x > a$, by that law which they are defined in an interval $(0, a)$.

Let's apply the Laplace integral transformation [6] to the differential equation (5.31), excluding derivative on $x$ replacing it with its linear expression concerning the image desired function. There are other functions for which the Laplace integral transformation converges absolutely. The valid part of the complex number $p = \alpha + i\beta, i = \sqrt{-1}$ it is considered positive, that is Re $p > 0$. We assume that the required solution $v(t, x)$, and its derivatives entering into the Eq. (5.31), satisfy Laplace integral transformation conditions on $x$, and its growth degree on $x$ function $v(t, x)$ and its derivatives do not depend on $t$. Multiplying both parts of the Eq. (5.31) on $\exp(-px)$ and integrating on $x$ from 0 to $\infty$, we have [6, 15], capital letters denoting images $V^{(1)}, H_1$ and $\frac{\partial g}{\partial x} = \frac{\partial v^{(1)}(t,0)}{\partial x}$, $g = v^{(1)}(t, 0)$. For simplicity, we omit the index calculations, $*$ down at $t$ and the index (1) above, at $v^{(1)}$, $T^{(1)}, g_i^{(1)}, q_i^{(1)}, i = 1, 2$ and below at $h_1, H_1$ then we have:

$$p^2 V(t,p) - pg - \frac{\partial g}{\partial x} = H(t,p) \quad \text{or}$$
$$V = \frac{g}{p} + \frac{\partial g/\partial x}{p^2} + \frac{H}{p^2}. \tag{5.34}$$

Using the return Laplace integral transformation [6]: $L^{-1}[1/p^2] = x$, $L^{-1}[H(p)/p] = \int_0^x h(y)\,dy$, let's restore the original for $v(t, x)$ of Eq. (5.34) [6]

$$v(t,x) = g + x\frac{\partial g}{\partial x} + \int_0^x (x-y)h(y)\,dy. \tag{5.35}$$

To find the unknown derivative $\partial g/\partial x$ and the function $g$ in Eq. (5.35) we differentiate the last $x$ [it is supposed that there is limited partial derivative on $x$ from $v(t, x)$]

$$\frac{\partial v}{\partial x} = \frac{\partial g}{\partial x} + \int_0^x h(y)\,dy. \tag{5.36}$$

Finally, the derivative $\partial g/\partial x$ and the function $g$ we receive, if we use Eqs. (5.33), (5.35), and (5.36) are:

$$\left.\frac{\partial v}{\partial x}\right|_{x=0} = q_1 - g_1 v|_{x=0} = \frac{\partial g}{\partial x}, \tag{5.37}$$

$$\left.\frac{\partial v}{\partial x}\right|_{x=a} = q_2 - g_2 v|_{x=a} = \frac{\partial g}{\partial x} + \int_0^a h(y)\,dy, \tag{5.38}$$

$$v(a) = g + a\frac{\partial g}{\partial x} + \int_0^a (a-y)h(y)\,dy. \tag{5.39}$$

Having substituted $v(a)$ of Eq. (5.39) in the Eq. (5.38) we find from the system (5.37)–(5.39):

$$g_1 g + \frac{\partial g}{\partial x} = q_1, \quad g_2 g + B_1\frac{\partial g}{\partial x} = B_2, \tag{5.40}$$

$$B_1 = 1 + a g_2, \quad B_2 = q_2 - \int_0^a h(y)[1 + g_2(a-y)]\,dy. \tag{5.41}$$

Further, solving the system of the Eqs. (5.40), we have

$$g = \frac{B_1 q_1 - B_2}{\Delta}, \quad \frac{\partial g}{\partial x} = \frac{B_2 g_1 - g_2 q_1}{\Delta}, \quad \Delta = B_1 g_1 - g_2. \tag{5.42}$$

Let's substitute $g$ and $\partial g/\partial x$ of Eq. (5.42) in Eq. (5.35) and given Eq. (5.41), we have, returning the index (1) at the top and $g_i^{(1)}, q_i^{(1)}, i = 1, 2$ 1 and the index at the bottom of the $h_1$:

$$v = S_1(t,x) - K_1(t,x)\int_0^a h_1(y)[1 + g_2^{(1)}(a-y)]\,dy$$

$$+ \int_0^x (x-y)h_1(y)\,dy, \tag{5.43}$$

$$S_1(t,x) = [q_1^{(1)}(1 + a g_2^{(1)}) - q_2^{(1)} + x(g_1^{(1)}q_2^{(1)} - g_2^{(1)}q_1^{(1)})]/\Delta,$$

$$K_1(t,x) = (x g_1^{(1)} - 1)/\Delta, \quad \Delta = g_1^{(1)}(1 + a g_2^{(1)}) - g_2^{(1)}. \tag{5.44}$$

We transform the expression on the right-hand side (5.43) to get rid of the second integral with a variable top limit. Then, entering Green's function $E(x,y)$ [7, 15, 16], we have

$$E(x,y) = \begin{cases} K_1[1 + g_2^{(1)}(a-y)] + y - x, & 0 \le y \le x, \\ K_1[1 + g_2^{(1)}(a-y)], & x \le y \le a, \end{cases} \tag{5.45}$$

$$v = S_1(t,x) - \int_0^a E(x,y)h_1(y)\,dy. \tag{5.46}$$

Then the expression (5.46), using formulas (5.30) and (5.31) can be rewritten by noting that $b_\xi, A(T)$ is clearly not depend on $x$ in contrast off the $R_\xi$ from Eq. (5.30):

$$b_\xi \frac{\partial}{\partial t}\left(\frac{A_H}{A}\frac{\partial v}{\partial t}\right)\int_0^a E(x,y)\,dy + v$$
$$= S_1 + \int_0^a E(x,y)R_\xi\,dy = Z_1, \quad 0 < t \leq t_*. \tag{5.47}$$

We apply again the Kirchhoff's transformation (5.19), to return to the original variable $T$ in Eq. (5.12)

$$T = \int_0^v \frac{A_H}{A}\,dv, \quad A = NT^m, \quad T = (v\phi)^{1/s}. \tag{5.48}$$

We transform the left-side of the Eq. (5.47), using Eq. (5.48):

$$\frac{\partial T}{\partial t} = \frac{A_H}{A}\frac{\partial v}{\partial t}, \quad \frac{\partial^2 T}{\partial t^2} = \frac{\partial}{\partial t}\left(\frac{A_H}{A}\frac{\partial v}{\partial t}\right), \quad v = \frac{T^s}{\phi}, \; \phi = \frac{sA_H}{N}, \; s = m+1.$$

Since the second term on the left-hand side of Eq. (5.47) takes the form $v = T^s/\phi$, then you must reapply method quasi-linearization (5.28), then have returning index top (1) and lower $*$

$$\frac{\partial^2 T^{(1)}}{\partial t^2} + \frac{s}{\phi}(T^{(0)})^m B_1 T^{(1)} = [X^{-1}Z_1 + (s-1)(T^{(0)})^s B_1/\phi] = P_1(t_*, x),$$
$$X = b_\xi \int_0^a E(x,y)\,dy, \quad B_1 = X^{-1}. \tag{5.49}$$

We use the Doetsch [4] applicability of the Laplace integral transformation to the partial differential equation derivatives as many times as its dimension. Then, having the initial conditions (5.32), we have

$$T^{(1)}(t_*, x) = s_1(t_*)p_1(x) + s_2(t_*)p_2(x)$$
$$+ \int_0^{t_*} s_2(t_* - \tau)P_1(\tau, x)\,d\tau. \tag{5.50}$$

$$s_1(t_*) = \cosh(\gamma t_*),\; s_2(t_*) = \gamma^{-1}\sinh(\gamma t_*),\; s_1(t_* - \tau) = \cosh[\gamma(t_* - \tau)],$$
$$s_2(t_* - \tau) = \gamma^{-1}\sinh[\gamma(t_* - \tau)],\; \beta < 0,\; \gamma = \sqrt{\beta},\; \beta = s(T^{(0)})^m B_1/\phi;$$
$$s_1(t_*) = \cos(\gamma t_*),\; s_2(t_*) = \gamma^{-1}\sin(\gamma t_*),\; s_1(t_* - \tau) = \cos[\gamma(t_* - \tau)],$$
$$s_2(t_* - \tau) = \gamma^{-1}\sin[\gamma(t_* - \tau)],\; \beta > 0.$$

$$\partial T^{(1)}(t_*, x)/\partial t) = s_1(t_*)p_2(x) + \mu s_2(t_*)p_1(x)$$
$$+ \int_0^{t_*} s_1(t_* - \tau)P_1(\tau, x)\, d\tau, \quad \mu = \text{sign}\,(\gamma^2) = \mp\gamma^2,$$
$$\partial v^{(1)}(t_*, x)/\partial t = N(T^{(1)})^m[\partial T^{(1)}(t_*, x)/\partial t]/A_H, \tag{5.51}$$

where the sign "minus" is taken at $\beta > 0$, and the "plus" is taken if $\beta < 0$ of Eq. (5.51).

We now have the analytical solution of the parabolic part (5.24)–(5.26) of the telegraph equation, using an algorithm (5.28)–(5.47). For simplicity, we omit writing the index $*$ down at $t$, the index (2) above from $v^{(2)}$, $T^{(2)}, g_i^{(2)}, q_i^{(2)}, i = 1, 2$ and the index (2) under from $h_2, H_2$, indicating capital letters image $V^{(2)}, H_2$, $\frac{\partial g}{\partial x} = \frac{\partial v^{(2)}(t_*,0)}{\partial x} = \frac{\partial v^{(1)}(t_*,0)}{\partial x}$, $g = v^{(2)}(t_*, 0) = v^{(1)}(t_*, 0)$, then we have:

$$p^2 V(t,p) - pg - \partial g/\partial x + \Phi_\eta V(t,p) + \gamma_\eta[pV(t,p) - g] = H(t,p)$$

or

$$V = \frac{(p+\delta)g}{(p+\delta)^2 + b^2} + \frac{b(\delta g + \partial g/\partial x + H)}{b[(p+\delta)^2 + b^2]}, \tag{5.52}$$

where $\delta = \gamma_\eta/2, b = \sqrt{\Phi_\eta - \delta^2}$. Using the inverse Laplace integral transformation [6]: $L^{-1}[H(p)/p] = \int_0^x h(y)\, dy$, $L^{-1}[(p+\delta)^{-1}] = \exp(-\delta x)$, we restore the original for $v(t, x)$ of Eq. (5.52). Then we have:

$$v(t,x) = \left[gw(x) + u_2(x)\frac{\partial g}{\partial x} + \int_0^x \exp(\delta y)u_2(x-y)h(y)\, dy\right]$$
$$\times \exp(-\delta x), \tag{5.53}$$

$w(x) = u_1(x) + \delta u_2(x), u_1(x) = \cos(bx), u_2(x) = b^{-1}\sin(bx), u_2(x - y) = b^{-1}\sin[b(x-y)]$, if $b^2 = \Phi_\eta - \delta^2 > 0$ from Eq. (5.30);

$$u_1(x) = \cosh(bx), \quad u_2(x) = b^{-1}\sinh(bx),$$
$$u_2(x-y) = b^{-1}\sinh[b(x-y)], \quad \text{if} \quad b^2 < 0. \tag{5.54}$$

To find the unknown derivative $\partial g/\partial x$ and the function $g$ in Eq. (5.53) we differentiate the last on $x$ (it is supposed that there is a limited partial derivative on $x$ from $v(t, x)$). Then we note at $\delta$ of Eqs. (5.52) and (5.30), clearly does not depend on $x$:

$$\frac{\partial v}{\partial x} = -\delta \exp(-\delta x)\left[gw(x) + u_2(x)\frac{\partial g}{\partial x} + \int_0^x \exp(\delta y)u_2(x-y)h(y)\,dy\right]$$
$$+ \exp(-\delta x)\left\{g[\mu u_2(x) + \delta u_1(x)] + u_1(x)\frac{\partial g}{\partial x} + \int_0^x \exp(\delta y)\right.$$
$$\left.\times\, u_1(x-y)h(y)\,dy\right\}, \quad \mu = \operatorname{sign} b^2 = \mp b^2. \tag{5.55}$$

Finally a derivative $\partial g/\partial x$ and a function $g$ give us, if we use Eqs. (5.26), (5.53), and (5.55):

$$\left.\frac{\partial v}{\partial x}\right|_{x=0} = q_1 - g_1 v|_{x=0} = \frac{\partial g}{\partial x}, \tag{5.56}$$

$$\left.\frac{\partial v}{\partial x}\right|_{x=a} = q_2 - g_2 v|_{x=a} = \exp(-\delta a)\left\{gw(a) + u_2(a)\frac{\partial g}{\partial x}\right.$$
$$+ \int_0^a \exp(\delta y)u_2(a-y)h(y)\,dy + r(a)\frac{\partial g}{\partial x}$$
$$\left.+ \int_0^a \exp(\delta y)r(a-y)h(y)\,dy - gu_2(a)(\delta^2 \pm b^2)\right\}, \tag{5.57}$$

$$r(a) = u_1(a) - \delta u_2(a), \quad r(a-y) = u_1(a-y) - \delta u_2(a-y),$$
$$v(a) = \exp(-\delta a)\left[gw(a) + u_2(a)\frac{\partial g}{\partial x}\right.$$
$$\left.+ \int_0^a \exp(\delta y)u_2(a-y)h(y)\,dy\right]. \tag{5.58}$$

Having substituted $v(a)$ of Eq. (5.58) in Eq. (5.57), we find

$$gg_1 + \frac{\partial g}{\partial x} = q_1, \quad gB_1 + \frac{\partial g}{\partial x}B_2 = B_3, \tag{5.59}$$

$$B_1 = \exp(-\delta a)[g_2 w(a) - u_2(a)(\delta^2 \pm b^2)],$$
$$B_2 = \exp(-\delta a)[r(a) + g_2 u_2(a)],$$
$$B_3 = q_2 - \int_0^a \exp[\delta(ya)][g_2 u_2(a-y) + r(a-y)]h(y)\,dy, \tag{5.60}$$

where the sign "minus" is taken in $b^2 = \Phi_\eta - \delta^2 < 0$, and "plus" is taken at $b^2 > 0$ in Eq. (5.54).

Further, solving the system of Eq. (5.59), we have

$$g = \frac{B_2 q_1 - B_3}{\Delta}, \quad \frac{\partial g}{\partial x} = \frac{B_3 g_1 - B_1 q_1}{\Delta}, \qquad \Delta = B_2 g_1 - B_1. \tag{5.61}$$

Let's substitute $g$ and $\partial g/\partial x$ of Eq. (5.61) in Eq. (5.53) and considering Eq. (5.60), returning the index (2) at the top and $g_i^{(2)}, q_i^{(2)}, i = 1, 2$ and the index 2 at the bottom of the $h_2$, we have:

$$v(t,x) = -M \int_0^a \exp[\delta(y-a)][r(a-y) + g_2^{(2)} u_2(a-y)] h_2(y)\, dy$$
$$+ \int_0^x \exp[\delta(y-x)] u_2(xy) h_2(y)\, dy + S_2(t,x), \tag{5.62}$$

$$M = \exp(-\delta x)[g_1^{(2)} u_2(x) - w(x)]/\Delta,$$
$$S_2(t,x) = \exp[-\delta(x+a)]\{w(x) q_1^{(2)} [g_2^{(2)} u_2(a) + r(a)] - q_2^{(2)}\}/\Delta$$
$$+ u_2(x) \exp[-\delta(x+a)]\{g_1^{(2)} q_2^{(2)} + q_1^{(2)} [u_2(a)(\delta^2 \mp b^2)$$
$$- g_2^{(2)} w(a)]\}/\Delta, \quad \Delta = \exp(-\delta a)\{g_1^{(2)} [g_2^{(2)} u_2(a) + r(a)]$$
$$- [g_2^{(2)} w(a) - u_2(a)(\delta^2 \pm b^2)]\}. \tag{5.63}$$

We transform the expression on the right-hand side of Eq. (5.62) to get rid of the second integral with a variable top limit. Then, entering Green's function $G(x,y)$ [7, 15, 16]

$$G(x,y) = \begin{cases} M \exp[\delta(y-a)][g_2^{(2)} u_2(a-y) + r(a-y)] \\ - \exp[\delta(y-x)] u_2(x-y), \quad 0 \le y \le x, \\ M \exp[\delta(y-a)][g_2^{(2)} u_2(a-y) + r(a-y)], \quad x \le y \le a, \end{cases} \tag{5.64}$$

expression (5.62), using formulas (5.30), (5.63), and (5.64) and returning the superscript (2) in top at $v$ and lower $*$ at $t$, it will be rewritten

$$\dot{v}^{(2)} + U v^{(2)} = Y^{-1} \left[ S_2(t_*, x) + \int_0^a G(x,y) R_\eta(v^{(1)})\, dy \right] = W(t_*, x),$$
$$Y = \int_0^a G(x,y) c_\eta(v^{(1)})\, dy, \quad U = Y^{-1}, \quad v^{(2)}(0,x) = v^{(1)}(t_*, x),$$
$$v^{(2)}(t_*, x) = v_{n+1}(t_*, x). \tag{5.65}$$

As a result the problem solution (5.24)–(5.26) will be [33]:

$$v_{n+1}(t_*,x) = \left[v^{(1)}(t_*,x) + \int_0^{t_*} W(v^{(1)},\tau)\exp(\tau U)\,d\tau\right]$$
$$\times \exp(-t_* U), \quad v_n = v^{(1)}(t_*,x), \quad n = 0,1,2,\ldots, \tag{5.66}$$

and the solution $T(t_*,x)$ of the initial boundary problem (5.6), (5.7), and (5.11) will agree with inversion formula (5.27):

$$T_{n+1}(t_*,x) = [v_{n+1}(t_*,x)\phi]^{1/s}, \quad n = 0,1,2,\ldots. \tag{5.67}$$

For simplicity of further analysis we will write solutions working formulas (5.6), (5.7), and (5.11) for boundary conditions of the first type of problem:

$$T|_{x=0} = G_1(t), \quad T|_{x=a} = Q_1(a,t), \tag{5.68}$$

where $G_1$, $Q_1$ is given as continuous function of time on the boundary $\Gamma$.

Using Kirchhoff's transformation (5.19) and the inversion formula (5.27), we have instead of Eq. (5.68),

$$v|_{x=0} = g_1, \quad v|_{x=a} = q_1, \tag{5.69}$$

where $g_1 = G_1^s/\phi$, $q_1 = Q_1^s/\phi$, and the initial condition (5.7) retains the same form.

Then formulas (5.45), (5.46), (5.63), and (5.64) are rewritten [15, 16]

$$S_1(t,x) = g_1 + (q_1 - g_1)x/a,$$

$$E(x,y) = \begin{cases} y(a-x)/a, & 0 \le y \le x, \\ x(a-y)/a, & x \le y \le a, \end{cases} \tag{5.70}$$

$$S_2(t,x) = \exp(-\delta x)\{q_1 w(x) + [q_1\exp(\delta a)$$
$$- g_1 w(a) u_2(x)/u_2(a)]\}, \tag{5.71}$$

$$G(x,y) = \begin{cases} \exp[\delta(y-x)][u_2(x)u_2(a-y)/u_2(a) - u_2(x-y)], & 0 \le y \le x, \\ \exp[\delta(y-x)]u_2(x)u_2(a-y)/u_2(a), & x \le y \le a. \end{cases} \tag{5.72}$$

For the boundary conditions of the second type $(A\partial T/\partial x)|_\Gamma = -S_t D|_\Gamma$ we have

$$E(x,y)=\begin{cases}(a-x), & 0\le y\le x,\\ (a-y), & x\le y\le a,\end{cases}$$

$$S_1(t,x)=(x-a)q_1-aq_2,\quad 0\le x<a,$$
$$q_i=-S_t\frac{D_i(\Gamma,t)}{A_H},\quad i=1,2,$$

$$T^{(1)}(t,a)=\frac{A_H}{N(T^{(0)})^m}\left[-aq_2+\left(\frac{s}{\phi}-1\right)(T^{(0)})^s\right]\Bigg|_{x=a}, \tag{5.73}$$
$$S_2(t,x)=\exp(-\delta x)\{q_2M(x-a)\exp(\delta a)+q_1[u_2(x)-u_2(a)M]\},$$
$$M=w(x)/w(a),$$
$$G(x,y)=\begin{cases}\exp[\delta(y-x)][Mu_2(a-y)-u_2(x-y)], & 0\le y\le x,\\ \exp[\delta(y-x)]Mu_2(a-y), & x\le y\le a.\end{cases}$$

$$v^{(2)}(t,a)=v_{n+1}(t,a)=S_2(t,a). \tag{5.74}$$

The form of the solutions (5.50), (5.51), (5.65), and (5.66) in the presence of the inversion formula (5.67) will stay the same. That is the genericity of the developed mathematical technology.

## Existence, Uniqueness, and Convergence

Without loss of generality, we will consider the solution of the simplified boundary value problem (5.6), (5.7), and (5.68) in $G_1=Q_1=0, p_2=0$.

We take a nonlinear case $m>0$, $k>1$ in the region $Q=(0<x<a)$, $\overline{Q}=Q+\Gamma$, $\overline{Q}_t=\overline{Q}\times[0<t\le t_0]$ for $A_2=0, C=1, A_1=-1, Y=0$, $A(T)=T^m$.

After applying the Kirchhoff's transformation (5.19) at $A_H=1$ and formulas (5.20) and (5.22) we have the modified boundary problem

$$b_\xi\frac{\partial}{\partial t}\left(\frac{1}{A}\frac{\partial v^{(1)}}{\partial t}\right)=\frac{\partial^2 v^{(1)}}{\partial x^2},\quad 0<t<t_*,$$
$$c_\eta\frac{\partial v^{(2)}}{\partial t}=\frac{\partial^2 v^{(2)}}{\partial x^2}-a_1(sv^{(2)})^{k/s},\quad 0<t<t_*,$$
$$T\,|_{t=0}=p_1(x),\quad (\partial T/\partial t)|_{t=0}=0,\quad v^{(i)}|_\Gamma=0,\quad i=1,2, \tag{5.75}$$

where $b_\xi=1/(\xi c^2), c_\eta=1/(\eta A), \phi=s, s=m+1$, $a_1=1/\eta$, and $v$ is determined from the inversion formula

$$v = T^s/s, \quad T = (vs)^{1/s}. \tag{5.76}$$

As a result of applying the algorithm (5.21)–(5.66) we have

$$v_{n+1}(t,x) = \left\{ v^{(1)}(t,x) + \int_0^t W(v^{(1)}, \tau) \exp[\tau U(v^{(1)})]\, d\tau \right\} \times \exp[-tU(v^{(1)})], \quad v_n = v^{(1)}(t,x), \quad n = 0, 1, 2, \ldots, \tag{5.77}$$

$$U = Y^{-1}, \quad Y = \int_0^a c_\eta(v^{(1)}) G(x,y)\, dy, \quad W = Y^{-1} \left[ S_2(t,x) + \int_0^a G(x,y) R_\eta(v^{(1)})\, dy \right], \quad v^{(1)} = (T^{(1)})^s/s, \quad T^{(1)} = p_1(x) s_1(t) + \int_0^t s_2(t-\tau) P_1(\tau, x)\, d\tau, \quad P_1 = X^{-1} Z_1, \quad Z_1 = S_1, \quad S_1 = 0,$$
$$A_2 = 0,$$

$$R_\eta = a_1 (Z_2)^{k/s} \left( 1 - \frac{v^{(1)} k}{Z_2} \right), \quad Z_2 = v^{(1)} s,$$
$$X = b_\xi \int_0^a E(x,y)\, dy, \quad s_1(t) = \cos(\gamma t), \quad \gamma = \sqrt{\beta},$$
$$\beta = (T^{(0)})^m B_1, \quad B_1 = X^{-1}, \tag{5.78}$$

where $E(x,y)$ and $G(x,y)$ are taken from formulas (5.70) and (5.72) with $\delta = 0, S_1 = 0, S_2 = 0$ according to the first formula (5.70) and the expression (5.71).

The final solution of the $T$ boundary value problem (5.6), (5.7), and (5.68) with $G_1 = Q_1 = 0, p_2 = 0$ is received after substitution $v_{n+1}$ of Eq. (5.77) in the inversion formula (5.76).

**Theorem.** *Let $T$ be continuously differentiated in $\overline{Q}_t$, then in region $\overline{Q}_t$ there is a unique solution to the problem* (5.6), (5.7), *and* (5.68).

Existence and uniqueness of solutions of a boundary problem (5.6), (5.7), and (5.68) is proved in the same manner as in Chapter 2 [see formulas (2.19)–(2.30)].

## Estimation of Speed of Convergence [15, 16]

It is considered that in some neighborhood of a root function $f = f(v, \dot{v})$ of Eq. (5.28), together with partial derivatives $\partial f/\partial v$, $\partial^2 f/\partial v^2$, $\partial f/\partial \dot{v}$, $\partial^2 f/\partial \dot{v}^2$

are continuous and $\partial f/\partial v$, $\partial^2 f/\partial v^2$, $\partial f/\partial \dot{v}$, $\partial^2 f/\partial \dot{v}^2$ in this neighborhood do not go to zero.

Let's address the recurrence relationship (5.28), noticing that $f(v, \dot{v}) = s(v)+r(\dot{v})$ in Eq. (5.28), we will subtract the $n$-e the equation from $(n+1)$th, then we will find:

$$\frac{\partial^2(v_{n+1} - v_n)}{\partial x^2} = s(v_n) - s(v_{n-1}) - (v_n - v_{n-1})\frac{\partial s(v_{n-1})}{\partial v} + (v_{n+1} - v_n)\frac{\partial s(v_n)}{\partial v} + \left[r(\dot{v}_n) - r(\dot{v}_{n-1}) - (\dot{v}_n - \dot{v}_{n-1})\frac{\partial r(\dot{v}_{n-1})}{\partial \dot{v}} + (\dot{v}_{n+1} - \dot{v}_n)\frac{\partial r(\dot{v}_n)}{\partial \dot{v}}\right]. \tag{5.79}$$

From the average theorem [33] it follows:

$$s(v_n) - s(v_{n-1}) - (v_n - v_{n-1})\frac{\partial s(v_{n-1})}{\partial v} = 0.5(v_n - v_{n-1})^2 \times \frac{\partial^2 s(\xi)}{\partial v^2}, \quad v_{n-1} \le \xi \le v_n.$$

Let's consider Eq. (5.79) how the equation is in relation to $u_{n+1} = v_{n+1} - v_n$, $(u_0 = v_n, u_n = v_n - v_{n-1})$ and transform it as above Eqs. (5.52), (5.53), (5.77), and (5.78). Then we will have:

$$u_{n+1} = \int_0^a G(y, x)\left\{\left[u_n^2\frac{\partial^2 s(v_n)}{\partial v^2} + \dot{u}_n^2\frac{\partial^2 r(\dot{v}_n)}{\partial \dot{v}^2}\right]/2 + u_{n+1}\frac{\partial s(v_n)}{\partial v} + \dot{u}_{n+1}\frac{\partial r(\dot{v}_n)}{\partial \dot{v}}\right\} dy$$

or

$$\dot{u}_{n+1} - u_{n+1}\left[1 - \int_0^a G(x, y)\frac{\partial s(v_n)}{\partial v}\,dy\right]Y^{-1} = -0.5Y^{-1}\int_0^a G(x, y) \times \left[u_n^2\frac{\partial^2 s(v)}{\partial v^2} + \dot{u}_n^2\frac{\partial^2 r(\dot{v}_n)}{\partial \dot{v}^2}\right] dy, \quad u_H = 0,$$

$$Y = \int_0^a G(x, y)\frac{\partial r(\dot{v}_n)}{\partial \dot{v}}\,dy; \tag{5.80}$$

$$G(x, y) = \begin{cases} y(x - a)/a, & 0 \le y \le x, \\ x(y - a)/a, & x \le y \le a. \end{cases}$$

Finally the problem solution (5.80) will look like Eq. (5.77), where $u_0 = v^{(1)}(t, x) = 0$.

Let's put $\max\limits_{v,\dot{v}\in V}\left(\left|\frac{\partial s(v)}{\partial v}\right|, \left|\frac{\partial r(\dot{v})}{\partial \dot{v}}\right|\right) = c_1$, $\max\limits_{v,\dot{v}\in V}\left(\left|\frac{\partial^2 s(v)}{\partial v^2}\right|, \left|\frac{\partial^2 r(\dot{v})}{\partial \dot{v}^2}\right|\right) = c_2$, $\max\limits_{x,y}|G(x, y)| = a/4$ [7] assuming $c_m < \infty, m = 1, 2$. Then from the Eqs. (5.77) and (5.80) it follows that $\frac{\partial^2 r(\dot{v})}{\partial \dot{v}^2} = 0$:

$$|u_{n+1}| \le B\exp(t\alpha)\int_0^t u_n^2 \exp(-\alpha\tau)\,d\tau. \tag{5.81}$$

Let's choose $u_0(t, x)$ so that $|u_0(t, x)| \le 1$ is in the region $\overline{Q}_t$. As a result from expression (5.81) at $n = 0$, introducing $M_1 = \max\limits_{\overline{Q}_t}|u_1|$, we find $z_0 = \max\limits_{\overline{Q}_t} u_0^2$, $z_0 \le 1$, $B = c_2/2c_1$, $Y = c_1 a^2/4$, $\alpha = 4/c_1 a^2 - 1$:

$$M_1 \le \frac{B[\exp(t\alpha) - 1]}{\alpha} = S. \tag{5.82}$$

Hence, under a condition $\alpha > 0(a < 2/\sqrt{c_1})$ we find that the top border $M_1$ will not surpass 1 if there is inequality $S \le 1$ in Eq. (5.82):

$$t \le \ln\left(\frac{\alpha}{B} + 1\right)^{1/\alpha}. \tag{5.83}$$

Therefore, if choosing an interval $[0, t]$, $[0, a]$ small enough so that it satisfies the condition (5.83), we will have $M_1 \le 1$. Finally, we receive definitively $M_{n+1} \le S z_n$ or

$$\max_{x,t\in\overline{Q}_t} |v_{n+1} - v_n| \le S \max_{x,t\in\overline{Q}_t} |v_n - v_{n-1}|^2. \tag{5.84}$$

The relationship (5.84) shows, that if the convergence of the iterative process for a boundary problem (5.6), (5.7), and (5.68) according to the inversion formula (5.76) in general takes place, it is quadratic. Thus, with a big enough $n$ each following step doubles a number of correct signs in the given approximation.

## Results of Test Checks

Estimation of error of the analytical formulas (5.27), (5.44)–(5.51), (5.64)–(5.67) are checked practically in the solution of a boundary problem for the equations in partial derivatives in the region $\overline{Q}_t : [0 \le x \le a, 0 < t \le t_0]$ :

$$A_3\frac{\partial T}{\partial t}+z\frac{\partial^2 T}{\partial t^2}=\frac{\partial}{\partial x}\left[A(T)\frac{\partial T}{\partial x}\right]+B_2\frac{\partial T}{\partial x}$$
$$+A_1T^k+F(x,t),\quad A=NT^m,\quad z=A_H/c^2,\qquad(5.85)$$

$$T|_{t=0}=\exp(y),\quad \left.\frac{\partial T}{\partial t}\right|_{t=0}=\frac{\exp(y)}{w},\quad y=\frac{x}{a},\qquad(5.86)$$

$$\left[A\frac{\partial T}{\partial x}+\alpha_1S_tT\right]_{x=0}=\alpha_1S_tD_1(t),$$
$$\left[A\frac{\partial T}{\partial x}+\alpha_2S_tT\right]_{x=a}=\alpha_2S_tD_2(t,a).\qquad(5.87)$$

We have taken the exact solution of a problem (5.85)–(5.87)

$$T=\exp(\tau+y),\quad \tau=t/w,\qquad(5.88)$$

then the sources $F, D_1, D_2$ in the Eqs. (5.85) and (5.87) will be

$$F=\exp(\tau+y)\left\{\frac{z}{w^2}+\frac{A_3}{w}-\frac{B_2}{a}-\frac{Ns\exp[m(\tau+y)]}{a^2}\right\}$$
$$-A_1\exp[k(\tau+y)],$$
$$\alpha_1S_tD_1=\exp(\tau)[Na^{-1}\exp(\tau m)+\alpha_1(1+\tau_r/w)],$$
$$\alpha_2S_tD_2=\exp(\tau+1)\left\{\frac{N}{a}\exp[(\tau+1)m]+\alpha_2(1+\tau_r/w)\right\}.$$

The following basic values of initial data were used: $\alpha_1 = -1$, $\alpha_2 = -1$, $w=1$, $A_H=N=0.0253$ W/(K · m), $t_0=10^{-4}$ s, $B_2=13$, $c=340$ m/s, $\xi=\eta=0.5$, $A_3=1.3\cdot 10^3$ J /(m$^3$ · K) (air environment [53]), $a=0.01$ m, $N_x=11, M_t=51, \Delta x=a/(N_x-1), \Delta t=t_0/(M_t-1)$ are number of checkouts steps on space and time at finding integrals in the Eqs. (5.45), (5.46), (5.50), (5.51), (5.60), and (5.66) by Simpson's formula [21].

The boundary problem (5.85)–(5.87) is solved by means of formulas (5.27), (5.44)–(5.51), (5.64)–(5.67). The number of iterations was traced [for total expressions of a type (5.66) and (5.77)] on relative change of an error vector:

$$||V_n||=\max_{x,t\in\overline{Q}_t}\left|\frac{v_{n+1}-v_n}{v_{n+1}}\right|.$$

**Table 5.1** A dependence of the maximum relative errors at the various values $A_1, m, k$

| | Results of calculations | | | |
|---|---|---|---|---|
| **Variant number** | $A_1$ | $m$ | $k$ | $\varepsilon$, % |
| 1 | 0 | 0 | 1 | 0.2 |
| 2 | −1 | 0 | 1 | 4.43 |
| 3 | 0 | 0.5 | 1 | 6.49 |
| 4 | 1 | 0.5 | 1 | 7.25 |
| 5 | 1 | 0.5 | 2 | 8.83 |
| 6 | −1 | 0.5 | 1 | 5.72 |
| 7 | −1 | 0.5 | 2 | 5.03 |
| 8 | −2 | 0.5 | 1 | 4.92 |
| 9 | −2 | 0.5 | 2 | 3.57 |

Table 5.1 shows the results of test calculations $||V_n|| \leq \delta, \delta = 0.01$. This took only two or three iterations to achieve this accuracy and computation time of any variant is $t_p = 1$ s.

The program is made in the language G-Fortran, calculation was made on Pentium (3.5 GHz) with double accuracy. Table 5.1 gives the maximum relative error

$$\varepsilon = \frac{|T - \tilde{T}|100\%}{T}, \tag{5.89}$$

where $T$ is the exact explicit solution of Eq. (5.88), $\tilde{T}$ is the approximate analytical solution on mathematical technology of this section at various values $k$, $A_1$, $m$.

As is seen from Table 5.1, calculation on developed mathematical technology has almost the small error $\varepsilon$ from Eq. (5.89).

For the boundary conditions of the second type $(A\partial T/\partial x)|_\Gamma = -S_t D|_\Gamma$ we have the formulas (5.49)–(5.51), (5.66), (5.67), (5.73), and (5.74) and this is reflected in Table 5.2 at $B_2 = 0.01$ and $t_0 = 5 \cdot 10^{-6}$ s.

At the boundary conditions of the first type:

$$T|_{x=0} = \exp(\tau), \quad T|_{x=a} = \exp(\tau + 1) \tag{5.90}$$

an approximate analytical solution of the problem (5.85), (5.86), and (5.90) was received by the formulas (5.46), (5.50), (5.51), (5.65)–(5.67), and (5.70) and is reflected in Table 5.3 at $t_0 = 5 \cdot 10^{-6}$ s and $m = 0.5$.

Now we will compare the accuracy of the analytical formulas (5.27), (5.46), (5.50), (5.51), (5.65)–(5.67), and (5.70) simplified for the first boundary problem (5.85), (5.86), and (5.90)

**Table 5.2** A dependence of the maximum relative errors at the various values $A_1, m, k$

| Variant number | Results of calculations $A_1$ | $m$ | $k$ | $\varepsilon, \%$ |
|---|---|---|---|---|
| 1 | 0 | 0 | 1 | 2.89 |
| 2 | 1 | 0 | 1 | 2.88 |
| 3 | −1 | 0.5 | 1 | 4.54 |
| 4 | 1 | 0.5 | 1 | 4.53 |
| 5 | 1 | 0.5 | 2 | 4.52 |
| 6 | 10 | 0.5 | 1 | 4.49 |
| 7 | 10 | 0.5 | 2 | 4.47 |
| 8 | −10 | 0.5 | 1 | 4.57 |
| 9 | −10 | 0.5 | 2 | 4.59 |

**Table 5.3** A dependence of the maximum relative errors at the various values $A_1, B_2, k$

| Variant number | Results of calculations $A_1$ | $B_2$ | $k$ | $\varepsilon, \%$ |
|---|---|---|---|---|
| 1 | 1 | 0.13 | 1 | 3.1 |
| 2 | 1 | −0.13 | 1 | 6.87 |
| 3 | 1 | −0.26 | 1 | 8.7 |
| 4 | 1 | 0.26 | 1 | 3.96 |
| 5 | 1 | 0.26 | 2 | 3.95 |
| 6 | −1 | −0.26 | 2 | 8.71 |
| 7 | −10 | −0.26 | 2 | 8.66 |
| 8 | −10 | 0.26 | 2 | 4.01 |
| 9 | 10 | −0.26 | 2 | 8.75 |

$$\chi^{-1}\frac{\partial T}{\partial t} + z^{-1}\frac{\partial^2 T}{\partial t^2} = \frac{\partial^2 T}{\partial y^2}, \quad 0 < t \le t_0, \tag{5.91}$$

$$T|_{t=0} = T_H, \quad (\partial T/\partial t)|_{t=0} = 0, \quad T|_{y=0} = T_w, \quad T|_{y\to\infty} = T_H \tag{5.92}$$

with the known analytical solution of [49] at $z = c^2$, $\chi = A_H/C$

$$T(y,t) = \theta(T_w - T_H), \quad \theta = u(\beta - \delta)\exp(-\delta) + \delta u(\beta - \delta)\int_0^{\sqrt{\beta^2-\delta^2}} \frac{I_1(\eta)\exp[-(\eta^2+\delta^2)^{0.5}]}{(\eta^2+\delta^2)^{0.5}}\,d\eta, \tag{5.93}$$

$$\eta = \sqrt{\beta^2 - \delta^2}, \quad \beta = c^2 t/(2\chi), \quad \delta = cy/(2\chi),$$

where $u(\beta - \delta)$ is unit function ($u = 1$ for $\beta > \delta$, $u = 0$ for $\beta < \delta$), $I_1(\eta)$ is modified function first order Bessel [49]

$$I_1 = \sum_{k=0}^{\infty} \frac{(\eta/2)^{2k+1}}{(k+1)(k!)^2}. \tag{5.94}$$

Formula (5.94) shows that series converges slowly at $\eta > 20$, so on formulas (5.93) and (5.94), for example, in metals: steel—$c = 5700\,\mathrm{m/s}$, $\chi = 5 \cdot 10^{-6}\,\mathrm{m^2/s}$, copper—$c = 4700\,\mathrm{m/s}$, $\chi = 1.16 \cdot 10^{-4}\,\mathrm{m^2/s}$, the liquid: water—$c = 1500\,\mathrm{m/s}$, $\chi = 1.43 \cdot 10^{-7}\,\mathrm{m^2/s}$ [53], air medium and other at normal conditions can realistically investigate only rapid processes ($t \leq 10^{-8}$ s) on microspace ($a \leq 10^{-6}$ m), described by one-dimensional problems (5.91) and (5.92).

At $\chi = 2 \cdot 10^{-5}\,\mathrm{m^2/s}$, $c = 340\,\mathrm{m/s}$, $T_w = 800\,\mathrm{K}$, $T_H = 293\,\mathrm{K}$, $a = 4 \cdot 10^{-7}$ m, $t_0 = 2.5 \cdot 10^{-9}$ s, $\beta = 7.225$, $0 \leq \delta \leq 3.4$, $0 \leq \eta \leq 6.375$, $\Delta\eta = 6.375/(N_1 - 1)$, $N_1 = 51$ unlike the analytical solution (5.27), (5.46), (5.50), (5.51), (5.65), (5.66), and (5.70) from the exact (5.93) did not exceed 5.6%.

## 5.2 METHOD OF SOLUTION OF THE THREE-DIMENSIONAL EQUATION IN PARTIAL DERIVATIVES

### Statement of a Problem and a Method Algorithm

Let us attempt to find a solution for the three-dimensional equation in the partial derivatives second order of hyperbolic type [47] with sources

$$z^{-1}\frac{\partial^2 w}{\partial t^2} + C(w)\frac{\partial w}{\partial t} = \sum_{j=1}^{3} \frac{\partial}{\partial x_j}\left[A(w)\frac{\partial w}{\partial x_j}\right]$$

$$+\sum_{j=1}^{3} Y_j(w)\frac{\partial w}{\partial x_j} + A_1 w^k + A_2(x,t) \tag{5.95}$$

in a parallelepiped $Q : [x = (x_1, x_2, x_3), (0 < x_j < L_j; 0 < L_j < \infty, j = 1, 2, 3)]$, $\overline{Q} = Q + \Gamma$, $\overline{Q}_t = \overline{Q} \times [0 < t \leq t_0]$, $\Gamma$ is boundary surface of ranges of definition $Q$ with the initial conditions

$$w|_{t=0} = p_1(x), \quad (\partial w/\partial t)|_{t=0} = p_2(x), \quad x = (x_1, x_2, x_3) \tag{5.96}$$

and to simplify the calculations with the boundary condition of the first kind

$$w|_{\Gamma} = \Psi, \quad \Psi \neq \text{const}, \tag{5.97}$$

where $A_1 = \text{const}, z = c^2/A_H, Y_i = -C(w)e_i(w), i = 1,2,3, A_H = \text{const}, c = \text{const}$, $c$ is velocity of propagation thermal disturbance (speed of sound in the medium); m/s, $e_i, i = 1,2,3$ is rate of convective heat-transfer, m/s; $C(w)$ is the coefficient of volume heat capacity, $\text{J}/(\text{K} \cdot \text{m}^3)$; $A(w)$ is coefficient of thermal conductivity, $A_H$ is coefficient conductivity at the initial temperature $\text{W}/(\text{K} \cdot \text{m})$; $w$ is temperature thermal conductivity for the hyperbolic equation, K; $x_i, i = 1,2,3$ is axes of a Cartesian coordinate system, $L_i, i = 1,2,3$ length of the sides of the parallelepiped, m; $t_0$ is end of the time interval, s.

Following [17, 47], Eq. (5.95) received under the assumption that $C(w)$, $Y(w)$ explicit does not depend on time $t$ and $e_i < 1\,\text{m/s}, i = 1,2,3$ neglect mixed derivative $\sum_{j=1}^{3} e_j \frac{\partial^2 w}{\partial t \partial x_j}$ compared with $\frac{\partial^2 w}{\partial t^2}$ in the left-hand side of Eq. (5.95). Furthermore, in an equation on the right-side are absent mixed derivatives of form $\sum_{j=1,i\neq j}^{3} \frac{\partial}{\partial x_j}[A(w)\frac{\partial w}{\partial x_i}]$ and the summand is $\tau_r \frac{\partial A_2}{\partial t}$, $\tau_r = \chi/c^2$ ($\tau_r$ is time relaxation, $\chi$ is thermal diffusivity, $\text{m}^2/\text{s}$), a value which for time is discussed below: $t > 100 \cdot \tau_r$ ($\tau_r \sim 10^{-9}$ s) is negligible.

Let's assume everywhere:

1. A problem (5.95)–(5.97) has a unique solution $w(x,t)$, which is continuously in the closed region $\overline{Q}_t$ and has continuous derivatives $\frac{\partial w}{\partial t}, \frac{\partial^2 w}{\partial t^2}, \frac{\partial w}{\partial x_j}, \frac{\partial^2 w}{\partial x_j^2}, j = 1,2,3$.
2. The following conditions are satisfied: $A(w) \geq l_1 > 0, C(w) \geq l_2 > 0$, $z \geq l_3 > 0$, $l_1, l_2, l_3$ are constants; $p_1, p_2$ are continuous defined functions in $\overline{Q}$, and $C, A, Y_j, A_2, j = 1,2,3$ are continuous function in the closed area $\overline{Q}_t$.
3. Coefficients $C(w), Y(w)$ in the general case can be nonlinear manner dependent on the problem solution [5], form $A(w)$ is defined below in a formula (5.110), and $\Psi$ is the given continuous function on $\Gamma$ for $0 < t \leq t_0$, having limited partial derivatives of the first order.

We apply locally-one-dimensional scheme decomposition to Eqs. (5.95)–(5.97) on a differential level [12] and introduce the superscripts (1), (2), (3) to designate the intermediate stages of solving the problem, as well as the direction of $\xi$ as a solution of a wave and a direction $\eta$ is the parabolic part a solution of the original equation (5.95). Then we have

$$z^{-1}\frac{\partial^2 w_\xi^{(1)}}{\partial t^2} = \xi\frac{\partial}{\partial x_1}\left[A(w_\xi^{(0)})\frac{\partial w_\xi^{(1)}}{\partial x_1}\right] + \xi\sigma_1 A_2, \quad 0 < t < t_*, \tag{5.98}$$

$$w_\xi^{(0)}|_{t=0} = p_1(x), \quad \left.\frac{\partial w_\xi^{(0)}}{\partial t}\right|_{t=0} = p_2(x), \quad 0 < x_j < L_j, \quad j = 1,2,3,$$

$$w_\xi^{(1)}|_{x_1=0} = G_1(t_*, x_2, x_3), \quad w_\xi^{(1)}|_{x_1=L_1} = G_2(t_*, L_1, x_2, x_3), \tag{5.99}$$

$$C(w_\xi^{(1)})\frac{\partial w_\eta^{(1)}}{\partial t} = \eta\frac{\partial}{\partial x_1}\left[A(w_\xi^{(1)})\frac{\partial w_\eta^{(1)}}{\partial x_1}\right] + Y_1(w_\xi^{(1)})\frac{\partial w_\eta^{(1)}}{\partial x_1}$$
$$+ \sigma_1[A_1(w_\eta^{(1)})^k + \eta A_2], \quad 0 < t < t_*, \tag{5.100}$$

$$w_\eta^{(1)}(0, x) = w_\xi^{(1)}(t_*, x), \quad w_\eta^{(1)}|_{x_1=0} = G_1, \quad w_\eta^{(1)}|_{x_1=L_1} = G_2; \tag{5.101}$$

$$z^{-1}\frac{\partial^2 w_\xi^{(2)}}{\partial t^2} = \xi\frac{\partial}{\partial x_2}\left[A(w_\eta^{(1)})\frac{\partial w_\xi^{(2)}}{\partial x_2}\right] + \xi\sigma_2 A_2, \quad 0 < t < t_*, \tag{5.102}$$

$$w_\xi^{(2)}|_{t=0} = w_\eta^{(1)}(t_*, x), \quad \left.\frac{\partial w_\xi^{(2)}}{\partial t}\right|_{t=0} = \frac{\partial w_\eta^{(1)}(t_*, x)}{\partial t}, \; 0 < x_j < L_j, \; j = 1,2,3,$$

$$w_\xi^{(2)}|_{x_2=0} = Q_1(t_*, x_1, x_3), \quad w_\xi^{(2)}|_{x_2=L_2} = Q_2(t_*, x_1, L_2, x_3), \tag{5.103}$$

$$C(w_\xi^{(2)})\frac{\partial w_\eta^{(2)}}{\partial t} = \eta\frac{\partial}{\partial x_2}\left[A(w_\xi^{(2)})\frac{\partial w_\eta^{(2)}}{\partial x_2}\right] + Y_2(w_\xi^{(2)})\frac{\partial w_\eta^{(2)}}{\partial x_2}$$
$$+ \sigma_2[A_1(w_\eta^{(2)})^k + \eta A_2], \quad 0 < t < t_*, \tag{5.104}$$

$$w_\eta^{(2)}(0, x) = w_\xi^{(2)}(t_*, x), \quad w_\eta^{(2)}|_{x_2=0} = Q_1, \quad w_\eta^{(2)}|_{x_2=L_2} = Q_2; \tag{5.105}$$

$$z^{-1}\frac{\partial^2 w_\xi^{(3)}}{\partial t^2} = \xi\frac{\partial}{\partial x_3}\left[A(w_\eta^{(2)})\frac{\partial w_\xi^{(3)}}{\partial x_3}\right] + \xi\sigma_3 A_2, \quad 0 < t < t_*, \tag{5.106}$$

$$w_\xi^{(3)}|_{t=0} = w_\eta^{(2)}(t_*, x), \quad \left.\frac{\partial w_\xi^{(3)}}{\partial t}\right|_{t=0} = \frac{\partial w_\eta^{(2)}(t_*, x)}{\partial t}, \; 0 < x_j < L_j, \; j = 1,2,3,$$

$$w_{\xi}^{(3)}|_{x_3=0} = D_1(t_*, x_1, x_2), \quad w_{\xi}^{(3)}|_{x_3=L_3} = D_2(t_*, x_1, x_2, L_3), \tag{5.107}$$

$$C(w_{\xi}^{(3)})\frac{\partial w_{\eta}^{(3)}}{\partial t} = \eta\frac{\partial}{\partial x_3}\left[A(w_{\xi}^{(3)})\frac{\partial w_{\eta}^{(3)}}{\partial x_3}\right] + Y_3(w_{\xi}^{(3)})\frac{\partial w_{\eta}^{(3)}}{\partial x_3}$$
$$+ \sigma_3[A_1(w_{\eta}^{(3)})^k + \eta A_2] \quad 0 < t < t_*, \tag{5.108}$$

$$w_{\eta}^{(3)}(0, x) = w_{\xi}^{(3)}(t_*, x), \quad w_{\eta}^{(3)}|_{x_3=0} = D_1, \quad w_{\eta}^{(3)}|_{x_3=L_3} = D_2, \tag{5.109}$$

where $\xi + \eta = 1$, $\sigma_1 + \sigma_2 + \sigma_3 = 1$. If $C(w) \neq 0$ is solved by a system of Eqs. (5.98)–(5.109) telegraph equation: friction loss (conductive medium), if $C(w) = 0, A_1 = 0, \eta = 0$ is solved by a system of Eqs. (5.98), (5.99), (5.102), (5.103), (5.106), and (5.107) wave equation: the absence of friction (decaying medium).

In accordance with the method of splitting [12, 13] we are taking the next model, the convection-conduction head-conductivity for conductivity for the hyperbolic equation at $A_1 = 0, A_2 = 0$. At the first stage of convection-conduction, head-conductivity is switched off in the directions of coordinates $x_2, x_3$ which is a problem consider Eqs. (5.98)–(5.101). Then we receive temperature distribution $T^{(1)}(t_*, x)$. Accepting it for intermediate, we switched off convection-conduction head-conductivity in directions of coordinates $x_1, x_3$ and solved the problem (5.102)–(5.105) Then for $t = t_*$, we have a distribution of temperature $T^{(2)}(t_*, x)$. Taking it again for intermediate, we switched off convection-conduction head-conductivity in the direction of coordinates $x_1, x_2$ and consider the problems (5.106)–(5.109). Finally, we found at $t = t_*$ a temperature $T^{(3)}(t_*, x)$, which coincides with the sought value of temperature $T(t_*, x)$. According to this model the process of convection-conduction head-conductivity is "stretched" on time and takes place during a period of time $3t_*$, instead of $t_*$ [12, 13]. This approach to multidimensional partial differential equations with constant coefficients is proposed and justified in [12, 13]. For the wave equation $C(T) = 0, A_1 = 0, \eta = 0$ excluded by the coordinate directions $x_1, x_2, x_3$ are wave propagation velocity, shear elasticity of the medium, etc.

However, before this system (5.98)–(5.109), we must apply Kirchhoff's transformation [5] and method quasi-linearization [7], which receive the differential equation with the constant coefficients, which can be solved using the Laplace integral transformation [6].

In the future, we use the inversion formula, $A(w)$ in Eq. (5.95) as [5]:

$$A(w) = Nw^m, \tag{5.110}$$

where $m \geq 0$, $N > 0$, $N$ is constant [5].

We use Kirchhoff's transformation [5]

$$v = \int_0^w \frac{A(w)}{A_H}\, dw. \tag{5.111}$$

Then, taking into account the relationships [5]:

$$\nabla A = \frac{\partial A}{\partial w}\nabla w, \quad \nabla v = \frac{A}{A_H}\nabla w, \quad \frac{\partial v}{\partial t} = \frac{A}{A_H}\frac{\partial w}{\partial t}, \tag{5.112}$$

we receive from Eqs. (5.98)–(5.112)

$$v = w^s/\phi, \quad \phi = sA_H/N, \quad s = m+1, \tag{5.113}$$

$$b\frac{\partial}{\partial t}\left(\frac{A_H}{A}\frac{\partial v_\xi^{(1)}}{\partial t}\right) = \frac{\partial^2 v_\xi^{(1)}}{\partial x_1^2} + \sigma_1 a_2, \quad 0 < t < t_*, \tag{5.114}$$

$$w_\xi^{(0)}|_{t=0} = p_1(x), \quad \left.\frac{\partial w_\xi^{(0)}}{\partial t}\right|_{t=0} = p_2(x), \quad 0 < x_j < L_j, \quad j = 1,2,3, \tag{5.115}$$

$$v_\xi^{(1)}|_{x_1=0} = g_1, \quad v_\xi^{(1)}|_{x_1=L_1} = g_2, \tag{5.116}$$

$$c_1\frac{\partial v_\eta^{(1)}}{\partial t} = \frac{\partial^2 v_\eta^{(1)}}{\partial x_1^2} + r_1\frac{\partial v_\eta^{(1)}}{\partial x_1} + \sigma_1[a_1(w_\eta^{(1)})^k + a_2], \quad 0 < t < t_*, \tag{5.117}$$

$$v_\eta^{(1)}(0,x) = v_\xi^{(1)}(t_*,x), \quad v_\eta^{(1)}|_{x_1=0} = g_1, \quad v_\eta^{(1)}|_{x_1=L_1} = g_2; \tag{5.118}$$

$$b\frac{\partial}{\partial t}\left(\frac{A_H}{A}\frac{\partial v_\xi^{(2)}}{\partial t}\right) = \frac{\partial^2 v_\xi^{(2)}}{\partial x_2^2} + \sigma_2 a_2, \quad 0 < t < t_*,$$

$$w_\xi^{(2)}|_{t=0} = w_\eta^{(1)}(t_*,x), \quad \left.\frac{\partial w_\xi^{(2)}}{\partial t}\right|_{t=0} = \frac{\partial w_\eta^{(1)}(t_*,x)}{\partial t},$$

$$v_\xi^{(2)}|_{x_2=0} = q_1, \quad v_\xi^{(2)}|_{x_2=L_2} = q_2,$$

$$c_2 \frac{\partial v_\eta^{(2)}}{\partial t} = \frac{\partial^2 v_\eta^{(2)}}{\partial x_2^2} + r_2 \frac{\partial v_\eta^{(2)}}{\partial x_2} + \sigma_2 [a_1 (w_\eta^{(2)})^k + a_2], \quad 0 < t < t_*,$$

$$v_\eta^{(2)}(0, x) = v_\xi^{(2)}(t_*, x), \quad v_\eta^{(2)}|_{x_2=0} = q_1, \quad v_\eta^{(2)}|_{x_2=L_2} = q_2; \qquad (5.119)$$

$$b \frac{\partial}{\partial t}\left( \frac{A_H}{A} \frac{\partial v_\xi^{(3)}}{\partial t} \right) = \frac{\partial^2 v_\xi^{(3)}}{\partial x_3^2} + \sigma_3 a_2, \quad 0 < t < t_*,$$

$$w_\xi^{(3)}|_{t=0} = w_\eta^{(2)}(t_*, x), \quad \left. \frac{\partial w_\xi^{(3)}}{\partial t} \right|_{t=0} = \frac{\partial w_\eta^{(2)}(t_*, x)}{\partial t},$$

$$v_\xi^{(3)}|_{x_3=0} = d_1, \quad v_\xi^{(3)}|_{x_3=L_3} = d_2,$$

$$c_3 \frac{\partial v_\eta^{(3)}}{\partial t} = \frac{\partial^2 v_\eta^{(3)}}{\partial x_3^2} + r_3 \frac{\partial v_\eta^{(3)}}{\partial x_3} + \sigma_3 [a_1 (w_\eta^{(3)})^k + a_2], \quad 0 < t < t_*,$$

$$v_\eta^{(3)}(0, x) = v_\xi^{(3)}(t_*, x), \quad v_\eta^{(3)}|_{x_3=0} = d_1, \quad v_\eta^{(3)}|_{x_3=L_3} = d_2, \qquad (5.120)$$

where $c_i = C(w_\xi^{(i)})/(\eta A), r_i = Y_i/(\eta A), i = 1, 2, 3, b = 1/(\xi c^2), a_1 = A_1/(\eta A_H), a_2 = A_2/A_H, g_i = G_i^s/\phi, q_i = Q_i^s/\phi, d_i = D_i^s/\phi, i = 1, 2$.

In this intermediate value in the directions: $w_\xi^{(j)}(t_*, x)$, $w_\eta^{(j)}(t_*, x)$, $j = 1, 2, 3$ are determined from Eq. (5.113) according to treatment formulas:

$$w_\xi^{(j)} = (\phi v_\xi^{(j)})^{1/s}, \quad w_\eta^{(j)} = (\phi v_\eta^{(j)})^{1/s}, \quad j = 1, 2, 3. \qquad (5.121)$$

Then finally the solution of Eqs. (5.95)–(5.97) is written as:

$$w(t_*, x) = w_\eta^{(3)})(t_*, x). \qquad (5.122)$$

Let's note that variation ranges of independent variables and type of the boundary conditions do not change in relation to Kirchhoff's transformation (5.111), and within the inversion formula (5.121) the boundary condition of the first kind passes into Dirichlet's condition.

Our purpose is to receive a solution of a nonlinear boundary problem, if it exists, as a limit of sequence of solutions of linear boundary tasks. For this we use the results in [7, 8]. Assume further that all the coordinate directions in space is equal. Let $v^{(1)}$ is some initial approximation [as an initial approximation it is expedient to take $v^{(1)}$ the first formula of

Eq. (5.115), taking into account the first equation (5.113)]. For analysis simplicity, we will consider the quasi-one-case and the sequence $v_n(t,x)$, defined recurrence relation [7] (the dot and bar at the top denote the partial derivative with respect to time and space):

$$\frac{\partial^2 v_{n+1}}{\partial y^2} = f + (v_{n+1} - v_n)\frac{\partial f}{\partial v_n} + (\dot{v}_{n+1} - \dot{v}_n)\frac{\partial f}{\partial \dot{v}_n}$$
$$+ (v'_{n+1} - v'_n)\frac{\partial f}{\partial v'_n}, \quad f = f(v_n, v'_n, \dot{v}_n), \tag{5.123}$$

$$v_{n+1}|_\Gamma = \Psi^s/\phi, \quad v_H = v_{n+1}(0,x), \quad n = 0,1,2,\ldots, \tag{5.124}$$

where $y$ is any of the coordinates $x_j, j = 1,2,3$ in Eq. (5.123). Then at $y = x_1$ other coordinates (5.123) and (5.124), $0 < x_j < L_j, j = 2,3$ are changed parametrically. On the remained coordinates while receiving expressions (5.123) there is a circular replacement of indexes, when instead of $y$ we substitute, respectively $x_2, x_3$. Notice that at solution of the three-dimensional boundary problem (5.95)–(5.97), if in the first coordinate direction $x_1$ as the initial iteration acts as $v_n = v^{(1)}$, then subsequent iterations $v_{n+1}$ will be received from final expression $v_{n+1}(t_*, x) = v^{(3)}(t_*, x)$ [see comment below to the formulas (5.152)–(5.157)]. Then in the quasi-one-version equation (5.123) and (5.124) can be rewritten on coordinate $x_1$ [16]:

$$\frac{\partial^2 v^{(1)}}{\partial x_1^2} = f_1 + (v^{(1)} - v^{(0)})\frac{\partial f_1}{\partial v^{(0)}} + (\dot{v}^{(1)} - \dot{v}^{(0)})\frac{\partial f_1}{\partial \dot{v}^0}$$
$$+ (v'^{(1)} - v'^{(0)})\frac{\partial f_1}{\partial v'^{(0)}}, \quad f_1 = f_1(v^{(0)}, v'^{(0)}, \dot{v}^{(0)}), \tag{5.125}$$

$$v^{(1)}(0,x) = v^{(0)}(t_*, x), \quad v^{(0)} = v_H,$$
$$v^{(1)}|_{x_1=0} = g_1, \quad v^{(1)}|_{x_1=L_1} = g_2. \tag{5.126}$$

Expressions similar to Eqs. (5.125) and (5.126) make it possible to write expressions like these on other coordinate directions $x_2, x_3$. In particular, for the second coordinate direction $x_2$ it is necessary in Eqs. (5.125) and (5.126) to replace the top and bottom indexes (1) and 1 on (2) and 2, and the top index (0) on (1). Thus for the initial condition in the second coordinate direction $x_2$ we have $v^{(2)}(0,x) = v^{(1)}(t_*, x)$.

Each function $v_{n+1}$ in Eqs. (5.123) and (5.124) in a quasi-one-case or $v^{(1)}$ in Eqs. (5.125) and (5.126) is a solution of the linear equation that is a rather important feature of this algorithm. The algorithm comes from an approximation of Newton-Kantorovich's method [8] in functional space.

To reduce further records we will introduce notations:

$$R_{\xi}^{(j)} = \sigma_j a_2, \quad f_j = c_j \dot{v}_{\eta}^{(j-1)} - r_j \frac{\partial v_{\eta}^{(j-1)}}{\partial x_j} - \sigma_j (a_1 Z_j^{k/s} + a_2),$$

$$c_j(v_{\eta}^{(j-1)}) = \frac{\partial f_j}{\partial \dot{v}_{\eta}^{(j-1)}}, \quad r_j = -\frac{\partial f_j}{\partial v_{\eta}^{\prime(j-1)}}, \quad \Phi_j = -\frac{\partial f_j}{\partial v_{\eta}^{(j-1)}},$$

$$\Phi_j = \frac{\sigma_j a_1 k Z_j^{k/s} A_H}{Z_j N}, \quad R_{\eta}^{(j)} = \sigma_j a_1 Z_j^{k/s} \left(1 - \frac{v_{\eta}^{(j-1)} k A_H}{Z_j N}\right)$$
$$+ \sigma_j a_2, \quad h_{\eta}^{(j)} = c_j \dot{v}_{\eta}^{(j-1)} - R_{\eta}^{(j)}, \quad Z_j = \phi v_{\eta}^{(j-1)}, \quad j = 1, 2, 3. \tag{5.127}$$

We receive the solution of a quasi-one problem (5.125) and (5.126) to coordinate direction $x_1$, using the Eqs. (5.114)–(5.116):

$$\frac{\partial^2 v_{\xi}^{(1)}}{\partial x_1^2} = h_1, \quad h_1 = b \frac{\partial}{\partial t}\left(\frac{A_H}{A} \frac{\partial v_{\xi}^{(1)}}{\partial t}\right) - R_{\xi}^{(1)}, \quad 0 < t < t_*, \tag{5.128}$$

$$w_{\xi}^{(0)}|_{t=0} = p_1(x), \quad \left.\frac{\partial w_{\xi}^{(0)}}{\partial t}\right|_{t=0} = p_2(x), \tag{5.129}$$

$$v_{\xi}^{(1)}|_{x_1=0} = g_1, \quad v_{\xi}^{(1)}|_{x_1=L_1} = g_2. \tag{5.130}$$

Let's apply the Laplace integral transformation to the differential equation (5.128), excluding derivative on respect to $x_1$ and replacing it with its linear expression concerning the image of desired function. We assume that the required solution $v^{(1)}(t, x)$ and its derivatives satisfy the Laplace integral transformation conditions on existence on $x_1$, and its growth degree on $x_1$ do not depend on $t, x_2, x_3$. We consider the functions for which the Laplace integral transformation is absolutely convergent. The valid part of complex number $p = \alpha + i\beta$, $i = \sqrt{-1}$ is positive, that is $\mathrm{Re}\, p > 0$. For simplicity for calculations, we omit the index $*$ bottom at $t$ and index $\xi$ at $v_{\xi}^{(1)}$, as well as the index (1) at the top and bottom $v_{\xi}^{(1)}$, $V^{(1)}$, $h_1 H_1$ , denoting $\partial g_1/\partial x_1 = \partial v_{\xi}^{(1)}(t, 0, x_2, x_3)/\partial x_1$, then we have [6]:

$$p^2V(t,p,x_2,x_3) - pg_1 - \frac{\partial g_1}{\partial x_1} = H(t,p,x_2,x_3), \quad 0 < x_j < L_j, \; j = 2,3 \text{ or}$$

$$V(t,p,x_2,x_3) = \frac{g_1}{p} + \frac{\partial g_1/\partial x_1}{p^2} + \frac{H(t,p,x_2,x_3)}{p^2}. \tag{5.131}$$

Using the return Laplace integral transformation [6]: $L^{-1}[1/p^2] = x_1$, $L^{-1}[H(p)/p] = \int_0^{x_1} h(y)\,dy$, we restore the original for $v(t,x)$ of Eq. (5.131) [6]

$$v(t,x) = g_1 + x_1\frac{\partial g_1}{\partial x_1} + \int_0^{x_1} (x_1 - y)h(y)\,dy. \tag{5.132}$$

Derivative $\partial g_1/\partial x_1$ in the expression (5.132) we find, using the second boundary condition the first coordinate direction $x_1$ of Eq. (5.130):

$$G_2 = g_1 + L_1\frac{\partial g_1}{\partial x_1} + \int_0^{L_1} (L_1 - y)h(y)\,dy. \tag{5.133}$$

Therefore, finding $\partial g_1/\partial x_1$ in Eq. (5.133) and substituting it in Eq. (5.132), we have

$$v(t,x) = g_1 + \frac{x_1}{L_1}\left[g_2 - g_1 - \int_0^{L_1} (L_1 - y)h(y)\,dy\right] + \int_0^{x_1} (x_1 - y)h(y)\,dy. \tag{5.134}$$

We transform the expression on the right side of Eq. (5.134) to get rid of the second integral with a variable upper limit. Then, entering Green's function $E(x_1, y)$ [16]

$$E_1(x_1,y) = \begin{cases} y(L_1 - x_1)/L_1, & 0 \le y \le x_1, \\ x_1(L_1 - y)/L_1, & x_1 \le y \le L_1, \end{cases} \tag{5.135}$$

expression (5.134) can be rewritten with returning superscript (1) and lower $\xi$, noting that $b, A(w)$ is clearly not depend on $x_1$ in contrast to the $R_\xi^{(1)}$ of Eq. (5.127):

$$b\frac{\partial}{\partial t}\left(\frac{A_H}{A}\frac{\partial v_\xi^{(1)}}{\partial t}\right)\int_0^{L_1} E(x_1,y)\,dy + v_\xi^{(1)} = g_1 + \frac{x_1}{L_1}(g_2 - g_1) + \int_0^{L_1} E(x_1,y)R_\xi^{(1)}\,dy = F_1, \quad 0 < t < t_*. \tag{5.136}$$

We apply again the Kirchhoff's transformation (5.111), to return to the original variable $w$ in Eq. (5.98)

$$w = \int_0^v \frac{A_H}{A}\, dv, \quad A = Nw^m, \quad w = (v\phi)^{1/s}. \tag{5.137}$$

We transform the left-side of the Eq. (5.136), using Eq. (5.137)

$$\frac{\partial w}{\partial t} = \frac{A_H}{A}\frac{\partial v}{\partial t}, \quad \frac{\partial^2 w}{\partial t^2} = \frac{\partial}{\partial t}\left(\frac{A_H}{A}\frac{\partial v}{\partial t}\right), \quad v = \frac{w^s}{\phi}, \quad \phi = \frac{sA_H}{N}.$$

Since the second term on the left-hand side of Eq. (5.136) takes a form $v_\xi^{(1)} = (w_\xi^{(1)})^s/\phi$, it is necessary to reapply method quasi-linearization (5.125) and (5.126), then we have

$$\frac{\partial^2 w_\xi^{(1)}}{\partial t^2} + \frac{s}{\phi}(w_\xi^{(0)})^m B_1 w_\xi^{(1)} = Y_1^{-1}F_1 + (s-1)(w_\xi^{(0)})^s B_1/\phi = P_1,$$

$$Y_1 = b\int_0^{L_1} E_1(x_1, y)\, dy, \tag{5.138}$$

where $B_1 = Y_1^{-1}$, $B_1 > 0$, $w_\xi^{(0)} = p_1(x)$, $E_1(x_1, y) > 0$ from Eq. (5.135).

We use the Doetsch [4] applicability of the Laplace integral transformation to the partial differential equation derivatives as many times as its dimension. Then, with the initial conditions (5.129), we receive from Eq. (5.138)

$$w_\xi^{(1)}(t_*, x) = s_1(t_*)p_1(x) + s_2(t_*)p_2(x) + \int_0^{t_*} s_2(t_* - \tau)P_1(\tau, x)\, d\tau, \tag{5.139}$$

$$\begin{aligned} s_1(t_*) &= \cos(\gamma_1 t_*), \quad s_2(t_*) = \gamma_1^{-1}\sin(\gamma_1 t_*), \quad s_1(t_* - \tau) \\ &= \cos[\gamma_1(t_* - \tau)], \quad s_2(t_* - \tau) = \gamma_1^{-1}\sin[\gamma_1(t_* - \tau)], \\ \gamma_1 &= \sqrt{\beta_1}, \quad \beta_1 = s(w_\xi^{(0)})^m B_1/\phi, \end{aligned} \tag{5.140}$$

$$\partial w_\xi^{(1)}(t_*, x)/\partial t = s_1(t_*)p_2(x) - \gamma_1^2 s_2(t_*)p_1(x) + \int_0^{t_*} s_1(t_* - \tau)P_1(\tau, x)\, d\tau. \tag{5.141}$$

We now have the analytical solution of the parabolic (5.117) and (5.118), using the notation (5.127):

$$\frac{\partial^2 v_\eta^{(1)}}{\partial x_1^2} + \Phi_1 v_\eta^{(1)} = h_\eta^{(1)} - r_1 \frac{\partial v_\eta^{(1)}}{\partial x_1}, \quad 0 < t < t_*, \tag{5.142}$$

$$v_\eta^{(1)}(0,x) = v_\xi^{(1)}(t_*,x), \quad v_\eta^{(1)}|_{x_1=0} = g_1, \quad v_\eta^{(1)}|_{x_1=L_1} = g_2.$$

Note that $\Phi_1$ of Eq. (5.127) clearly does not depend on $x_1$ in Eq. (5.142) [$\Phi_1$ can always be set at the lower iterate through $n$, knowing the meaning of $w_\xi^{(1)}$ in the initial and subsequent times of Eq. (5.139) and $v_\xi^{(1)}$ from the formula (5.113)]. For simplicity we omit writing the index $*$ down at $t$ and index $\eta$ and $v_\eta^{(1)}$, as well as the index of (1) at the top and bottom $v_\eta^{(1)}, V^{(1)}, h_\eta^{(1)}, H_1$. Then from Eqs. (5.128)–(5.135) we have

$$p^2 V(t,p,x_2,x_3) - pg_1 - \frac{\partial g_1}{\partial x_1} + \Phi_1 V(t,p,x_2,x_3)$$

$$+ r_1[pV(t,p,x_2,x_3) - g_1] = H(t,p,x_2,x_3), \quad 0 < x_j < L_j, \quad j = 2,3,$$

$$V = \frac{(p+\delta)g_1}{(p+\delta)^2 + b_1^2} + \frac{b_1(\delta g_1 + \partial g_1/\partial x_1 + H)}{b_1[(p+\delta)^2 + b_1^2]}, \tag{5.143}$$

where $\delta = r_1/2, b_1 = \sqrt{\Phi_1 - \delta^2}$.

Using the return Laplace integral transformation [6]: $L^{-1}[p/(p^2+b_1^2)] = \cos(b_1 x_1)$ at $b_1^2 = \Phi_1 - \delta^2 > 0$, $L^{-1}[p/(p^2 - b_1^2)] = \cosh(b_1 x_1)$ for $b_1^2 < 0$, $L^{-1}[H(p)/p] = \int_0^{x_1} h(y)\, dy$, $L^{-1}[(p+\delta)^{-1}] = \exp(-\delta x_1)$, we restore the original for $v(t,x)$ of Eq. (5.143) [6]

$$v(t,x) = \exp(-\delta x_1)\left\{ g_1[u_1(x_1) + \delta u_2(x_1)] + u_2(x_1)\frac{\partial g_1}{\partial x_1} \right.$$

$$\left. + \int_0^{x_1} \exp(\delta y) u_2(x_1 - y) h(y)\, dy \right\}, \quad 0 < x_j < L_j, \quad j = 2,3, \tag{5.144}$$

$$u_1(x_1) = \cos(b_1 x_1), \quad u_2(x_1) = b_1^{-1}\sin(b_1 x_1), \quad u_2(x_1 - y)$$

$$= b_1^{-1}\sin[b_1(x_1 - y)], \quad b_1^2 = \Phi_1 - \delta^2 > 0;$$

$$u_1(x_1) = \cosh(b_1 x_1), \quad u_2(x_1) = b_1^{-1}\sinh(b_1 x_1), \quad u_2(x_1 - y)$$

$$= b_1^{-1}\sinh[b_1(x_1 - y)], \quad b_1^2 < 0. \tag{5.145}$$

Derivative $\partial g_1/\partial x_1$ in the expression (5.144) we find using the second boundary condition on the first coordinate direction $x_1$ of Eq. (5.142), which has form

$$g_2(t, L_1, x_2, x_3) = \exp(-\delta L_1)\{g_1[u_1(L_1) + \delta u_2(L_1)] + u_2(L_1)\frac{\partial g_1}{\partial x_1} + \int_0^{L_1} \exp(\delta y)u_2(L_1 - y)h\,dy\}, \quad 0 < x_j < L_j, \quad j = 2, 3. \quad (5.146)$$

Therefore, finding $\partial g_1/\partial x_1$ in the expression (5.146) and substituting it into the Eq. (5.144), we find for $v$:

$$v(t, x) = S_1(t, x) + \int_0^{x_1} \exp[\delta(y - x_1)]u_2(x_1 - y)h\,dy - \frac{u_2(x_1)}{u_2(L_1)} \times \int_0^{L_1} \exp[\delta(y - x_1)]u_2(L_1 - y)h\,dy, \quad 0 < x_j < L_j, \quad j = 2, 3. \quad (5.147)$$

$$S_1(t, x) = \exp(-\delta x_1)g_1[u_1(x_1) + \delta u_2(x_1)] + \exp(-\delta x_1) \times \frac{u_2(x_1)}{u_2(L_1)}\{g_2 \exp(\delta L_1) - g_1[u_1(L_1) + \delta u_2(L_1)]\}. \quad (5.148)$$

We transform the expression on the right side of Eq. (5.147) to get rid of the first integral with a variable upper limit. Then, by introducing the Green function $G_1(x_1, y)$ [16]

$$G_1(x_1, y) = \begin{cases} \exp[\delta(y - x_1)][u_2(x_1)u_2(L_1 - y) \times u_2^{-1}(L_1) - u_2(x_1 - y)], \\ 0 \le y \le x_1, \\ \exp[\delta(y - x_1)]u_2(x_1)u_2(L_1 - y)/u_2(L_1), \quad x_1 \le y \le L_1, \end{cases} \quad (5.149)$$

expression (5.147), using formulas (5.127) and return lower index $*, \eta$, as well as the top—(1) can be rewritten as:

$$\dot{v}_\eta^{(1)} + U_1 v_\eta^{(1)} = X_1^{-1}\left[S_1(t_*, x) + \int_0^{L_1} G_1(x_1, y)R_\eta^{(1)}\,dy\right] = W_1(t_*, x),$$

$$X_1 = \int_0^{L_1} G_1(x_1, y)c_1(v_\xi^{(1)})\,dy, \quad U_1 = X_1^{-1}v_\eta^{(1)}(0, x) = v_\xi^{(1)}(t_*, x),$$

$$0 < x_j < L_j, \quad j = 2, 3. \quad (5.150)$$

As a result, the final solution of the problem (5.150) takes the form [33]

$$v_\eta^{(1)}(t_*, x) = \left[ v_\xi^{(1)}(t_*, x) + \int_0^{t_*} W_1(x, \tau) \exp(\tau U_1)\, d\tau \right]$$
$$\times \exp(-t_* U_1), \quad 0 < x_j < L_j, \quad j = 2, 3, \tag{5.151}$$

where $v_\xi^{(1)}(t_*, x)$ is taken from the formula (5.139) using Eq. (5.113).

We have a similar problem with Eqs. (5.119) and (5.120) to coordinate directions $x_2$ and $x_3$ using relations (5.121)–(5.127). Then on an algorithm (5.131)–(5.151) at $\partial^2 w_\eta^{(i)}(t_*, x)/\partial t^2 = 0, i = 1, 2, 3$ in Eqs. (5.100), (5.104), and (5.108) (these derivatives were not present therein) we have

$$w_\xi^{(i)}(t_*, x) = s_1(t_*) w_\eta^{(i-1)}(t_*, x) + s_2(t_*) \frac{\partial w_\eta^{(i-1)}(t_*, x)}{\partial t}$$
$$+ \int_0^{t_*} s_2(t_* - \tau) P_i(\tau, x)\, d\tau, \tag{5.152}$$
$$P_i = [Y_i^{-1} F_i + (s-1)(w_\eta^{(i-1)})^s B_i/\phi], \quad i = 2, 3,$$

$$w_\xi^{(i)}(t_*, x) = s_1(t_*) w_\xi^{(i-1)}(t_*, x) + s_2(t_*) \frac{\partial w_\xi^{(i-1)}(t_*, x)}{\partial t}$$
$$+ \int_0^{t_*} s_2(t_* - \tau) P_i(\tau, x)\, d\tau, \quad i = 2, 3, \tag{5.153}$$
$$P_i = [Y_i^{-1} F_i + (s-1)(w_\xi^{(i-1)})^s B_i/\phi],$$
$$Y_i = b \int_0^{L_i} E_i(x_i, y)\, dy, \quad B_i = Y_i^{-1},$$
$$F_2 = q_1 + \frac{x_2}{L_2}(q_2 - q_1) + \int_0^{L_2} E_2(x_2, y) R_\xi^{(2)}\, dy,$$
$$F_3 = d_1 + \frac{x_3}{L_3}(d_2 - d_1) + \int_0^{L_3} E_3(x_3, y) R_\xi^{(3)}\, dy,$$

$$\frac{\partial w_\xi^{(i)}(t_*, x)}{\partial t} = 2 s_1(t_*) \frac{\partial w_\eta^{(i-1)}(t_*, x)}{\partial t} - \gamma_i^2 s_2(t_*) \frac{\partial w_\eta^{(i-1)}(t_*, x)}{\partial t}$$
$$+ \int_0^{t_*} s_1(t_* - \tau) P_i(\tau, x)\, d\tau, \tag{5.154}$$
$$P_i = [Y_i^{-1} F_i + (s-1)(w_\eta^{(i-1)})^s B_i/\phi], \quad i = 2, 3,$$

$$\frac{\partial w_\xi^{(i)}(t_*, x)}{\partial t} = 2 s_1(t_*) \frac{\partial w_\xi^{(i-1)}(t_*, x)}{\partial t} - \gamma_i^2 s_2(t_*) \frac{\partial w_\xi^{(i-1)}(t_*, x)}{\partial t}$$
$$+ s_2(t_*) B_{i-1} \left[ F_{i-1} - \frac{(w_\xi^{(i-1)})^s}{\phi} \right] + \int_0^{t_*} s_1(t_* - \tau) P_i(\tau, x)\, d\tau, \tag{5.155}$$

$$P_i = [Y_i^{-1}F_i + (s-1)(w_\xi^{(i-1)})^s B_i/\phi], \quad i = 2, 3,$$

$$\frac{\partial w_\eta^{(i)}(t_*, x)}{\partial t} = s_1(t_*)\frac{A_H(w_\eta^{(i-1)})^{-m}}{N}\frac{\partial v_\eta^{(i-1)}(t_*, x)}{\partial t}, \tag{5.156}$$

$$\frac{\partial v_\eta^{(i-1)}(t_*, x)}{\partial t} = W_{i-1}(t_*, x) - U_{i-1}v_\eta^{(i-1)}(t_*, x), \quad i = 2, 3,$$

$$v_\eta^{(i)}(t_*, x) = \left[v_\xi^{(i)}(t_*, x) + \int_0^{t_*} W_i(x, \tau)\exp(\tau U_i)\, d\tau\right]$$
$$\times \exp(-t_* U_i), \quad i = 2, 3, \tag{5.157}$$

$$W_i = X_i^{-1}\left[S_i(t_*, x) + \int_0^{L_i} G_i(x_i, y)R_\eta^{(i)}\, dy\right], \quad U_i = X_i^{-1},$$

$$X_i = \int_0^{L_i} G_i(x_i, y)c_i(v_\xi^{(i)})\, dy, \; v_\eta^{(i)}(0, x) = v_\xi^{(i)}(t_*, x), \; i = 2, 3,$$

where formulas (5.152), (5.154), (5.156), and (5.157) are used for solving the telegraph equation and formulas (5.153) and (5.155) are used for solving wave equation ($C(w) = 0, A_1 = 0, \eta = 0$). In this $s_1(t_*), s_2(t_*)$, $s_2(t_* - \tau), s_1(t_* - \tau)$ is taken from expressions in Eq. (5.140) with a replacement index 1 at $\gamma_1, \beta_1, B_1$ appearing anywhere on the index 2 and 3.

$E_i(x_i, y), G_i(x_i, y), i = 2, 3$ are received from Eqs. (5.135) and (5.149), and $u_1(x_i), u_2(x_i), u_2(x_i - y), i = 2, 3$ from Eq. (5.145) by replacing all arguments $x_1, L_1$ respectively in order of appearing on $x_j, L_j, j = 2, 3$. $S_i(t_*, x), i = 2, 3$ from Eq. (5.148) are received similarly a replacement of $g_1, g_2$ to $q_1, q_2$, $d_1, d_2$ and corrections everywhere arguments $x_1, L_1$ respectively in order to follow $x_j, L_j, j = 2, 3$.

At $x = x_2$ in Eqs. (5.152)–(5.157) other variables $0 < x_j < L_j$, $j = 1, 3$ change parametrically as in Eq. (5.151). A similar situation for $x = x_3$, wherein we find the final solution (5.122) of the telegraph equation (5.95) with the boundary conditions (5.96) and (5.97) according to inversion formula (5.121): $w_{n+1}(t_*, x) = w_\eta^{(3)}(t_*, x)$, for any $t_* > 0, n = 0, 1, 2, \ldots$.

As can be seen from the algorithm, first intermediate values are excluded $w_\xi^{(j)}, v_\xi^{(j)}, w_\eta^{(j)}, v_\eta^{(j)}, j = 1, 2, 3$ of the Eqs. (5.113), (5.121), (5.139)–(5.141), (5.151)–(5.157) and formed formulas (5.152)–(5.157) for $v_{n+1}(t_*, x) = v_\eta^{(3)}(t_*, x)$, then from the inversion formula (5.121) is formed final expression (5.122): $w_{n+1}(t_*, x) = w_\eta^{(3)}(t_*, x)$, and then is included in the iterative process $n = 0, 1, 2, \ldots$.

## Existence, Uniqueness, and Convergence

We consider the nonlinear case $m > 0, k \geq 2$ and loss of generality with zero boundary conditions, the first kind in the cube $Q : (0 \leq x_j \leq b, b = \max(L_j), j = 1, 2, 3), 0 < t \leq t_0$ with $Y_j = 0, j = 1, 2, 3, A_1 = -1, A(w) = Nw^m, z = c^2/A_H$. Then a solution of the boundary problem

$$z^{-1}\frac{\partial^2 w}{\partial t^2} + C(w)\frac{\partial w}{\partial t} = \sum_{j=1}^{3} \frac{\partial}{\partial x_j}\left[A(w)\frac{\partial w}{\partial x_j}\right] - w^k, \quad 0 < t < t_0, \tag{5.158}$$

$$w(0, x) = p_1(x), \quad \dot{w}(0, x) = p_2, \quad w|_\Gamma = 0 \tag{5.159}$$

on an algorithm (5.125)–(5.157) with the notations:

$$c_j = 1/(\eta A), \quad \Phi_j = -\sigma_j k Z_j^{k/s} A_H/(Z_j N), \quad R_\eta^{(j)} = \sigma_j Z_j^{k/s} \times [v_\eta^{(j-1)} k A_H/(Z_j N) - 1]/(\eta A_H), \quad \sigma_j = 1/3, \quad Z_j = \phi v_\eta^{(j-1)}, \tag{5.160}$$

$$a = 1/(\xi c^2), \quad s = m + 1, \quad \delta = 0, \quad j = 1, 2, 3$$

written as:

$$v_\eta^{(3)}(t_*, x) = \left[v_\xi^{(3)}(t_*, x) + \int_0^{t_*} W_3(x, \tau)\exp(\tau U_3)\, d\tau\right]\exp(-t_* U_3), \tag{5.161}$$

$$w_\xi^{(3)}(t_*, x) = s_1(t_*)w_\eta^{(2)}(t_*, x) + s_2(t_*)\frac{\partial w_\eta^{(2)}(t_*, x)}{\partial t} + \int_0^{t_*} s_2(t_* - \tau)P_3(\tau, x)\, d\tau,$$

$$s_1(t_*) = \cos(\gamma_3 t_*), \quad s_2(t_*) = \gamma_3^{-1}\sin(\gamma_3 t_*), \quad s_1(t_* - \tau) = \cos[\gamma_3(t_* - \tau)], \quad s_2(t_* - \tau) = \gamma_3^{-1}\sin[\gamma_3(t_* - \tau)],$$
$$\gamma_3 = \sqrt{\beta_3}, \quad \beta_3 = s(w_\xi^{(2)})^m B_3/\phi,$$

$$W_3 = X_3^{-1}\int_0^{L_3} G_3(x_3, y)R_\eta^{(3)}\, dy, \quad X_3 = \int_0^{L_3} G_3(x_3, y)c_3\, dy,$$

$$P_3 = (s - 1)(w_\eta^{(2)})^s B_3/\phi, \quad B_3 = Y_3^{-1}, \quad U_3 = X_3^{-1},$$

$$Y_3 = a\int_0^{L_3} E_3(x_3, y)\,dy, \quad b_3 = \sqrt{\Phi_3}, \quad v_\eta^{(2)} = (w_\eta^{(2)})^s/\phi,$$

$u_2(x_3) = b_3^{-1}\sinh(b_3x_3)$, $u_2(x_3 - y) = b_3^{-1}\sinh[b_3(x_3 - y)]$ in (5.145),

$$\frac{\partial w_\eta^{(2)}(t_*, x)}{\partial t} = \frac{A_H(w_\eta^{(2)})^{-m}}{N}\frac{\partial v_\eta^{(2)}(t_*, x)}{\partial t},$$

$$\partial v_\eta^{(2)}(t_*, x)/\partial t = w_2(t_*, x) - U_2 v_\eta^{(2)}(t_*, x), \quad U_2 = X_2^{-1},$$
$$W_2 = X_2^{-1}\int_0^{L_2} G_2(x_2, y)R_\eta^{(2)}\,dy, \quad X_2 = \int_0^{L_2} G_2(x_2, y)c_2\,dy,$$

$$v_\eta^{(2)}(t_*, x) = \left[v_\xi^{(2)}(t_*, x) + \int_0^{t_*} W_2(x, \tau)\exp(\tau U_2)\,d\tau\right]\exp(-t_* U_2), \tag{5.162}$$

$$w_\xi^{(2)}(t_*, x) = \int_0^{t_*} s_2(t_* - \tau)P_2(\tau, x)\,d\tau + s_2(t_*)\frac{\partial w_\eta^{(1)}(t_*, x)}{\partial t} + s_1(t_*)w_\eta^{(1)}(t_*, x),$$

$$s_1(t_*) = \cos(\gamma_2 t_*), \quad s_2(t_*) = \gamma_2^{-1}\sin(\gamma_2 t_*), \quad s_1(t_* - \tau) = \cos[\gamma_2(t_* - \tau)], \quad s_2(t_* - \tau) = \gamma_2^{-1}\sin[\gamma_2(t_* - \tau)],$$
$$\gamma_2 = \sqrt{\beta_2}, \quad \beta_2 = s(w_\xi^{(1)})^m B_2/\phi,$$

$$P_2 = (s - 1)(w_\eta^{(1)})^s B_2/\phi, \quad B_2 = Y_2^{-1},$$
$$Y_2 = a\int_0^{L_2} E_2(x_2, y)\,dy, \quad b_2 = \sqrt{\Phi_2}, \quad v_\eta^{(1)} = (w_\eta^{(1)})^s/\phi,$$

$u_2(x_2) = b_2^{-1}\sinh(b_2x_2)$, $u_2(x_2 - y) = b_2^{-1}\sinh[b_2(x_2 - y)]$ in (5.145),

$$v_\eta^{(1)}(t_*, x) = \left[v_\xi^{(1)}(t_*, x) + \int_0^{t_*} W_1(x, \tau)\exp(\tau U_1)\,d\tau\right]\exp(-t_* U_1), \tag{5.163}$$

$$\partial v_\eta^{(1)}(t_*, x)/\partial t = w_1(t_*, x) - U_1 v_\eta^{(1)}(t_*, x), \quad U_1 = X_1^{-1},$$
$$W_1 = X_1^{-1}\int_0^{L_1} G_1(x_1, y)R_\eta^{(1)}\,dy, \quad X_1 = \int_0^{L_1} G_1(x_1, y)c_1\,dy,$$

$$w_\xi^{(1)}(t_*, x) = s_1(t_*)p_1(x) + s_2(t_*)p_2(x) + \int_0^{t_*} s_2(t_* - \tau)P_1(\tau, x)\,d\tau,$$

$$\begin{aligned} s_1(t_*) &= \cos(\gamma_1 t_*), \quad s_2(t_*) = \gamma_1^{-1}\sin(\gamma_1 t_*), \quad s_1(t_* - \tau) \\ &= \cos[\gamma_1(t_* - \tau)], \quad s_2(t_* - \tau) = \gamma_1^{-1}\sin[\gamma_1(t_* - \tau)], \\ \gamma_1 &= \sqrt{\beta_1}, \quad \beta_1 = s(w_\xi^{(0)})^m B_1/\phi, \end{aligned}$$

$$\begin{aligned} P_1 &= (s-1)(w_\eta^{(0)})^s B_1/\phi, \quad B_1 = Y_1^{-1}, \\ Y_1 &= a\int_0^{L_1} E_1(x_1, y)\,dy, \quad b_1 = \sqrt{\Phi_1}, \end{aligned}$$

$u_2(x_1) = b_1^{-1}\sinh(b_1 x_1)$, $u_2(x_1 - y) = b_1^{-1}\sinh[b_1(x_1 - y)]$ in (5.145), where $E_i(x_i, y), i = 2, 3$ is received from Eq. (5.135), replacing arguments $x_1, L_1$, respectively, in order to follow $x_j, L_j, j = 2, 3$.

It is considered that in some neighborhood of a root function $f = f(v, \dot{v})$ of Eq. (5.123), together with the partial derivatives $\partial f/\partial v$, $\partial^2 f/\partial v^2$, $\partial f/\partial \dot{v}$, $\partial^2 f/\partial \dot{v}^2$ are continuous and $\partial f/\partial v$, $\partial^2 f/\partial v^2$, $\partial f/\partial \dot{v}$, $\partial^2 f/\partial \dot{v}^2$ in this neighborhood do not go to zero.

**Theorem.** *Let $v$ be continuously differentiated in $\overline{Q}_t$, then in $\overline{Q}_t$ there is the unique solution of the boundary problem* (5.158) *and* (5.159).

Existence and uniqueness of solutions of a boundary problem (5.158) and (5.159) in the presence of the inversion formula (5.113) is proved in the same manner as shown in Chapter 2 [see formulas (2.19)–(2.30)].

## Estimation of Speed of Convergence

Let's turn to the recurrence relation (5.123) or (5.125), noticing that at $Y_i = 0, i = 1, 2, 3$ of Eq. (5.95) $f(v, \dot{v}) = s(v) + r(\dot{v})$ in Eq. (5.158), we will subtract the $n$-e equation from $(n+1)$th, which corresponds in the quasi-one version to the first equation (5.125) for $v^{(1)}$, then we receive the coordinate direction $x_1$

$$\begin{aligned} \partial^2(v^{(1)} - v_n)/\partial x_1^2 &= s(v_n) - s(v_{n-1}) - (v_n - v_{n-1})\partial s(v_{n-1})/\partial v \\ &+ (v^{(1)} - v_n)\partial s(v_n)/\partial v + r(\dot{v}_n) - r(\dot{v}_{n-1}) - (\dot{v}_n - \dot{v}_{n-1}) \\ &\times \partial r(\dot{v}_{n-1})/\partial \dot{v} + (\dot{v}^{(1)} - \dot{v}_n)\partial r(\dot{v}_n)/\partial \dot{v}. \end{aligned} \tag{5.164}$$

From the average theorem [33] it follows:

$$s(v_n) - s(v_{n-1}) - (v_n - v_{n-1})\frac{\partial s(v_{n-1})}{\partial v} = 0.5(v_n - v_{n-1})^2\frac{\partial^2 s(\xi)}{\partial v^2},$$

$$v_{n-1} \le \xi \le v_n.$$

Let's consider Eq. (5.164) how the equation is in relation $u^{(1)} = v^{(1)} - v_n, (u^{(0)} = u_n, u_n = v_n - v_{n-1})$ and transform it as above Eqs. (5.127)–(5.157). It is assumed that $v_n = (w_\xi^{(1)})^s/\phi$ is known from solutions (5.163) and system of Eqs. (5.160)–(5.163). Then we have:

$$u^{(1)} = \int_0^b E_1(y, x_1)\left\{\left[u_n^2\frac{\partial^2 s(v_n)}{\partial v^2} + \dot{u}_n^2\frac{\partial^2 r(\dot{v}_n)}{\partial \dot{v}^2}\right]/2 + u^{(1)}\frac{\partial s(v_n)}{\partial v} + \dot{u}^{(1)}\frac{\partial r(\dot{v}_n)}{\partial \dot{v}}\right\} dy$$

or

$$\dot{u}^{(1)} - u^{(1)}\left[1 - \int_0^b E_1(x_1, y)\left(\frac{\partial s(v_n)}{\partial v}\right) dy\right] Z_1^{-1} = -0.5 Z_1^{-1} \times \int_0^b E_1(x_1, y)\left[u_n^2\frac{\partial^2 s(v_n)}{\partial v^2} + \dot{u}_n^2\frac{\partial^2 r(\dot{v}_n)}{\partial \dot{v}^2}\right] dy,$$
$$u^{(0)} = 0, \quad Z_1 = \int_0^b E_1(x_1, y)\left[\frac{\partial r(\dot{v}_n)}{\partial \dot{v}}\right] dy, \quad \dot{u}_n = \dot{u}^{(0)}. \quad (5.165)$$

As a result the solution of Eq. (5.165) in the first coordinate direction $x_1$ will look like Eq. (5.151), where $u^{(0)} = 0$. Similarly, we can find a solution of form (5.157) to coordinate directions $x_2, x_3$ for the boundary problem (5.158) and (5.159). Finally, the solution like Eqs. (5.152)–(5.157), using formulas (5.162) and (5.163), can be rewritten at $i = 3$:

$$u^{(3)}(x, t_*) = \exp(t_* U)\int_0^{t_*} V_1 \exp(-\tau U_1)\, d\tau + \exp[t_*(U_2 + U_3)] \times \int_0^{t_*} V_2 \exp(-\tau U_2)\, d\tau + \exp(t_* U_3)\int_0^{t_*} V_3 \exp(-\tau U_3)\, d\tau, \quad (5.166)$$

$$U = \sum_{j=1}^{3} U_j(v^{(j-1)}), \quad U_j(v^{(j-1)}) = Z_j^{-1}\left\{1 - \int_0^b E_j(x_j, y) \times \left[\frac{\partial s(v^{(j-1)})}{\partial v}\right] dy\right\}, \quad V_j = -0.5 Z_j^{-1}\int_0^b E_j(x_j, y)\left[(u^{(j-1)})^2 \times \frac{\partial^2 s(v^{(j-1)})}{\partial v^2} + (\dot{u}^{(j-1)})^2\frac{\partial^2 r(\dot{v}^{(j-1)})}{\partial \dot{v}^2}\right] dy, \quad u^{(0)} = u_n,$$

$$Z_j = \int_0^b E_j(x_j, y)\left[\frac{\partial r(\dot{v}^{(j-1)})}{\partial \dot{v}}\right] dy, \quad j = 1, 2, 3, \quad n = 0, 1, 2, \ldots.$$

Let's put $\max\limits_{v,\dot{v}\in R}\left(\left|\frac{\partial s(v^{(j)})}{\partial v}\right|,\left|\frac{\partial r(\dot{v}^{(j)})}{\partial \dot{v}}\right|\right) = c_1, \max\limits_{x_j,y}|E_j(x_j,y)| = b/4$ [7], $\max\limits_{v,\dot{v}\in R}\left(\left|\frac{\partial^2 s(v^{(j)})}{\partial v^2}\right|,\left|\frac{\partial^2 r(\dot{v}^{(j)})}{\partial \dot{v}^2}\right|\right) = c_2, j = 0,1,2$, assuming $c_p < \infty, p = 1,2$. Then, noting that $\frac{\partial^2 r(\dot{v}^{(j)})}{\partial \dot{v}^2} = 0, j = 0,1,2$ and using an assumption of the equality of all directions in space ($U_j = \alpha$, $V_j = Bu_n^2, j = 1,2,3$) and an equality of functions $u^{(0)} = u^{(j)}, j = 1,2$ (for convergent sequence $v_n$ all intermediate values of $u^{(j)}, j = 0,1,2$ are close to zero as they are in a convergence interval: $[v^{(0)}, v^{(3)}]$), we have from Eqs. (5.165) and (5.166) at $u_{n+1}(t_*,x) = u^{(3)}(t_*,x)$, omitting an index $(*)$ in $t$ at the bottom:

$$|u_{n+1}| \leq Bu_n^2 V \exp(\alpha t)\int_0^t \exp(-\alpha\tau)\,d\tau,$$

$$V = \exp(2\alpha t) + \exp(\alpha t) + 1, \quad B = c_2/2c_1, \quad \alpha = 4/(c_1 b^2) - 1,$$
$$|u_{n+1}| \leq Bu_n^2[\exp(\alpha t) - 1]V/\alpha. \tag{5.167}$$

Let's choose $u_0(t,x)$ so that $|u_0(t,x)| \leq 1$ in $\overline{Q}_t$. As a result from the expression (5.167) at $n = 0$ we receive introducing $M_1 = \max\limits_{\overline{Q}_t}|u_1|$:

$$M_1 \leq \frac{B[\exp(3\alpha t) - 1]}{\alpha} = S. \tag{5.168}$$

Hence, under condition $\alpha > 0 (b < 2/\sqrt{c_1})$, we find that the top border $M_1$ will not surpass 1, if there is inequality $S \leq 1$ in Eq. (5.168):

$$t \leq \ln\left(\frac{\alpha}{B} + 1\right)^{1/3\alpha}. \tag{5.169}$$

Therefore choosing intervals $[0,t], [0,b]$ small enough so that a condition is satisfied from Eq. (5.169), we will have $M_1 \leq 1$. Finally we find by induction:

$$\max_{x,t\in\overline{Q}_t}|v_{n+1} - v_n| \leq S\max_{x,t\in\overline{Q}_t}|v_n - v_{n-1}|^2. \tag{5.170}$$

The relationship (5.170) shows that if a convergence of the iterative process for a boundary problem (5.158) and (5.159) according to the inversion formula (5.113) in generally takes place, it is quadratic. Thus, a big enough $n$ following step doubles a number of correct signs in the given approximation.

## Results of Test Checks

Estimation of error of the analytical formulas (5.113), (5.121), (5.139)–(5.141), (5.151)–(5.157) will be checked practically in solving a boundary problem for the equations in partial derivatives in the $\overline{Q}_t : [0 \le x_j \le L_j, j = 1, 2, 3] \times [0 < t \le t_0]$ at $A(w) = Nw^m, m > 0$:

$$z^{-1}\frac{\partial^2 w}{\partial t^2} + A_3\frac{\partial w}{\partial t} = \sum_{j=1}^{3}\frac{\partial}{\partial x_j}\left[A(w)\frac{\partial w}{\partial x_j}\right] + \sum_{j=1}^{3} A_2\frac{\partial w}{\partial x_j}$$

$$+ A_1 w^k + F(x, t), \quad 0 < t \le t_0, \tag{5.171}$$

$$w|_{t=0} = \exp(y), \ \frac{\partial w}{\partial t}|_{t=0} = \frac{\exp(y)}{b}, \quad b > 0, \quad b = \text{const},$$

$$y = \sum_{j=1}^{3} z_j, \quad z_j = x_j/L_j, \quad j = 1, 2, 3, \quad z = c^2/A_H, \tag{5.172}$$

$$w|_{x_1=0} = \exp(\tau + z_2 + z_3), \quad w|_{x_2=0} = \exp(\tau + z_1 + z_3),$$

$$w|_{x_3=0} = \exp(\tau + z_1 + z_2), \quad w|_{x_1=L_1} = \exp(\tau + 1 + z_2 + z_3),$$

$$w|_{x_2=L_2} = \exp(\tau + z_1 + 1 + z_3), \quad w|_{x_3=L_3}$$

$$= \exp(\tau + z_1 + z_2 + 1), \quad \tau = t/b. \tag{5.173}$$

We have taken the exact solution of a problem (5.171)–(5.173):

$$w = \exp(\tau + y), \tag{5.174}$$

then a source $F$ in an Eq. (5.171) will be

$$F = \exp(\tau + y)\left\{\frac{1}{zb^2} + \frac{A_3}{b} - A_2(L_1^{-1} + L_2^{-1} + L_3^{-1})\right.$$

$$\left. -s(L_1^{-2} + L_2^{-2} + L_3^{-2})N\exp[m(\tau + y)]\right\} - A_1\exp[k(\tau + y)].$$

The following basic values of initial data were used: $b = 1, \xi = \eta = 0.5$, $A_3 = 1.3 \cdot 10^3$ J/( K $\cdot$ m$^3$), $A_H = N = 0.0253$ W/(K $\cdot$ m), $c = 340$ m/s (air environment [53]), $t_0 = 10^{-3}$ s, $m = 0.5, \sigma_1 = 0.4, \sigma_2 = 0.3, \sigma_3 = 0.3$, $L_j = 0.1$ m, $j = 1, 2, 3; \Delta x_j = L_j/(N_j - 1), N_j = 11, j = 1, 2, 3$, $\Delta t = t_0/(M - 1), M = 51$ are number of checkouts steps on space and time while finding integrals in the Eqs. (5.136), (5.139)–(5.141), (5.150)–(5.157) by Simpson's formula [33].

The boundary problem (5.171)–(5.173) is solved here by means of formulas (5.113), (5.121), (5.139)–(5.141), (5.151)–(5.157). The number of iterations was traced [for total expressions of a type (5.152)–(5.157)] at $i = 3$ on relative change of an error vector:

$$||V_n|| = \max_{x,t\in\overline{Q}_t} \frac{|v_{n+1} - v_n|}{v_{n+1}}.$$

Table 5.4 shows the results of test calculations $||V_n|| \leq \delta, \delta = 10^{-2}$. A time of a calculation of any variant is $t_p = 2$ s.

The program is written in G-Fortran, calculation was made on Pentium (3.5 GHz) with double accuracy. In Table 5.4 the maximum relative error is given

$$\epsilon = \max_{x,t\in\overline{Q}_t} \frac{|w - \tilde{w}|100\%}{w}, \tag{5.175}$$

where $w$ is the exact explicit solution (5.174), $\tilde{w}$ is the approximate analytical solution to the mathematical techniques in this chapter at different values of $k$, $A_j, j = 1, 2$ for support variant. As seen from Table 5.4 the accuracy of the analytical solutions was received satisfactorily.

In comparison with the numerical solution of a problem (5.171)–(5.173) we used technology for the calculation of the linear telegraph equation of an article [54]. For the numerical calculation, implicity applied unconditionally stable difference scheme with absolute error of approximation for the first and second derivative in space was used $O\left[\sum_{j=1}^{3} h_j^2\right]$ and the three-level scheme for the time derivative with approximation error is

**Table 5.4** A dependence of the maximum relative error solving the telegraph equation

| | The calculation results | | | |
|---|---|---|---|---|
| Version number | $A_1$ | $A_2$ | $k$ | $\varepsilon$, % |
| 1 | 1 | −6.5 | 1 | 0.085 |
| 2 | 1 | 13 | 1 | 0.242 |
| 3 | 1 | −13 | 1 | 0.083 |
| 4 | 1 | 13 | 2 | 0.297 |
| 5 | 2 | 13 | 2 | 0.419 |
| 6 | −1 | 13 | 2 | 0.224 |
| 7 | −2 | 13 | 2 | 0.273 |
| 8 | −2 | −13 | 2 | 0.083 |

**Table 5.5** A dependence of the maximum relative error solution of the wave equation

| Version number | Results of calculations | | | |
|---|---|---|---|---|
| | $t_0$ | $m$ | $a$ | $\varepsilon, \%$ |
| 1 | $5 \cdot 10^{-6}$ | 0.5 | 0.1 | 4.29 |
| 2 | $10^{-4}$ | 0 | 0.1 | 5.03 |
| 3 | $5 \cdot 10^{-5}$ | 0.5 | 1 | 4.28 |
| 4 | $10^{-3}$ | 0 | 1 | 4.96 |

$O[(\Delta t)^2]$. For basic initial sizes $A_1 = m = 0$, $A_3 = 1300, t_0 = 0.2$ s, $A_2 = 1.3, L_i = 0.1$ m, $i = 1, 2, 3$ on Eqs. (5.113), (5.121), (5.139)–(5.141), (5.151)–(5.157) receive $\epsilon = 9.1\%$, and on the difference schemes [54] at $N_i = 41, i = 1, 2, 3, \Delta t = 10^{-3}$ is $\epsilon = 18.6\%$ and $t_p = 3$ s. The specific example is an accuracy of the analytical formulas (5.113), (5.121), (5.139)–(5.141), (5.151)–(5.157) had double the accuracy of the numerical solutions [54].

Table 5.5 shows the results of solving the wave equation using the formulas (5.139)–(5.141), (5.153), (5.155) for different values of $m, t_0, L_i = a, i = 1, 2, 3$. As seen from Table 5.5 a solution of the wave equation satisfactory accuracy ($\epsilon < 10\%$) is achieved for the small times $5 \cdot 10^{-6} \leq t_0 \leq 10^{-3}$ s.

## CONCLUSION

1. Based on locally-one-dimensional scheme splitting, method, quasi-linearization, and the Laplace integral transformation, the approximate analytical solution of nonlinear hyperbolic equations of second order type is found in theory of a row without using [19].
2. In the one-dimensional case comparison is given of the analytical accuracy formulas derived in this chapter, with the known exact solution of the telegraph equation [49].
3. The method of test function gives the results of comparison of analytical solutions, developed by technology, with the exact solution of the boundary problem and numerical solution, by a known method.
4. The test calculation problems (5.171)–(5.173) are considered as conventional finite spatiotemporal: $0.1 \leq L_i \leq 1$ m, $i = 1, 2, 3$, $10^{-5} \leq t_0 \leq 0.2$ s found on practice [17, 18, 47].

# CONCLUSION

On the basis of the aforementioned results, it is possible to conclude the following:

1. Combining known algorithms in the correct manner: Kirchhoff's transformation, methods of quasi-linearization, operational calculation, and splittings, it is possible to solve, analytically (precisely or approach), the equations in private derivatives of the first and second order of required spatial dimension.
2. For linear and nonlinear boundary problems, exact and approximate analytical solutions have been formulated. In cases of nonlinear boundary problems, the conditions of unequivocal resolvability have been determined and an estimation of speed of convergence of the iterative process has been made.
3. Test calculations of modeling boundary problems on the basis of trial functions have been demonstrated and compared with numerical methods.
4. It was possible to extend given mathematical technology for the solution of the conjugate problem of heat-exchange of two ideally adjoining bodies [55, 56] and for solving the telegraph equation.
5. It is expedient to develop this mathematical technology for the solution of nonlinear equations in partial derivatives of the second order in curvilinear orthogonal coordinates. This was demonstrated to show it is possible to establish a method of operational calculation suitable for the solution of the ordinary differential equations with constant factors of any order [4, 6].

# BIBLIOGRAPHY


[1] Marchuk GI. Mathematical modeling in a problems of environment. Moscow: Science; 1982. 319 p.
[2] Grishin AM, Fomin VM. Conjugate and non-stationary problems of mechanics of reacting environments. Novosibirsk: Science; 1984. 318 p.
[3] Gardner MF, Berns DL. Transients in linear systems with concentrated constants. Moscow: Physics Mathematical Publishing; 1961. 551 p.
[4] Dech G. The manual to practical application of transformations of Laplace. Moscow: Physics Mathematical Publishing; 1960. 370 p.
[5] Lykov AV. Methods of solution of nonlinear equations of non-stationary heat conductivity. In: Proceeding of academic the USSR. Power and transport, vol. 5; 1970. p. 109–50.
[6] Ditkin VA, Prudnikov AP. Operational calculation. Moscow: The Higher School; 1966. 406 p.
[7] Bellman P, Kalaba P. Quasi-linearization and nonlinear boundary problems. Moscow: World; 1968. 183 p.
[8] Kantorovich LV. The functional analysis and applied mathematics. Successes Math Sci 1948;3(6):89–185.
[9] Grishin AM. About one modification of M.E. Shvetsa's method. Eng Phys Mag 1970;19(1):84–93.
[10] Vajnberg AM, Kantorovich VK, Hiterer R. The solution of non-stationary heat and mass transfer in nonlinear Newton-Kantorovich's method. Theor Bases Chem Technol 1991;(6):805–13.
[11] Grishin AM, Bertsun VN, Zinchenko VI. The iterative-interpolation method and its appendices. Tomsk: Publishing House of Tomsk State University; 1981. 160 p.
[12] Samarskii AA. Introduction to the theory of difference schemes. Moscow: Science; 1971. 552 p.
[13] Samarskij AA. About numerical method of solution of problems of mathematical physics. Heat Mass Transfer 1969;11:990–1006.
[14] Kovenja VM, Janenko N. A splitting method in problems of gas dynamics. Novosibirsk: Science; 1981. 304 p.
[15] Yakimov AS. The analytical method of solution of boundary problems. Tomsk: Publishing House of Tomsk State University; 2005. 108 p.
[16] Yakimov AS. The analytical method for solving of mathematical physics some equations. Tomsk: Publishing House of Tomsk State University; 2007. 150 p.
[17] Lykov AV. Theory of heat conductivity. Moscow: The Higher School; 1967. 599 p.
[18] Tihonov AN, Samarskii AA. The equations of mathematical physics. Moscow: Science; 1977. 735 p.
[19] Fihtengolts GM. The course differential and integral calculations, vol. 2. Moscow: Physics Mathematical Publishing; 1962. 807 p.
[20] Kalitkin NN. Numerical methods. Moscow: Science; 1978. 512 p.
[21] Demidovich BP, Maron IA. Fundamentals of computing mathematics. Moscow: Science; 1966. 644 p.
[22] Grishin AM, Yakimov AS. About one method of solution of some three-dimensional equations in private derivatives. Comput Technol 2000;5(5):38–52.
[23] Samarskii AA, Nikolaev ES. Methods of solution of net equations. Moscow: Science; 1978. 591 p.
[24] Marchuk GI. Calculus mathematics methods. Moscow: Science; 1989. 603 p.

[25] Ilyin VP. Finite difference methods of solution of elliptic equations. Novosibirsk: Science; 1970. 263 p.
[26] Matveev NM. Methods of integration of ordinary differential equations. Moscow: The Higher School; 1967. 564 p.
[27] Ladyzhensky OA, Uraltseva NN. The linear and quasi-linear equations of elliptic type. Moscow: Science; 1973. 576 p.
[28] Vladimirov VS. The equations of mathematical physics. Moscow: Science; 1976. 527 p.
[29] Grishin AM, Yakimov AS. About one method of solution three-dimensional elliptic equation of a general view. Comput Technol 2001;6(2):73–83.
[30] Bronstein I, Semendjaev LA. The directory on mathematician for engineers and students of technical colleges. Moscow: Science. Main Edition Physical and Mathematical Literatures; 1986. 544 p.
[31] Polezhaev JV, Jurevich FB. Thermal protection. Moscow: Energy; 1976. 391 p.
[32] Yakimov AS. One method of solution of nonlinear transfer equation. Sib Mag Ind Math 2003;6(1):154–62.
[33] Vygodsky M. The directory on higher mathematics. Moscow: Physics Mathematical Publishing; 1962. 870 p.
[34] Yakimov AS. About one method of solution of linear transfer equation. Comput Technol 1995;4(10):322–32.
[35] Godunov SK, Zabrodin AV, Ivanov MY, Kraiko AN, Prokopov GP. Numerical solution of multidimensional problems of gas dynamics. Moscow: Science; 1976. 400 p.
[36] Yakimov AS, Kataev AG. Method of solution of three-dimensional nonlinear transfer equation. Sib Mag Ind Math 2004;7(2):148–61.
[37] Kartashov EM. Analytical methods in the theory of heat conductivity of the firm bodies. Moscow: The Higher School; 1985. 480 p.
[38] Natanson IP. Theory of functions of the material variable. Moscow: Science. The Main Edition of the Physical and Mathematical Literature; 1974. 480 p.
[39] Ditkin VA, Prudnikov AP. Operational calculation on two variables and its appendices. Moscow: Physics Mathematical Publishing; 1958. 178 p.
[40] Koshljakov NS, Gliner EB, Smirnov MM. The equations in private derivatives of mathematical physics. Moscow: Science; 1970. 710 p.
[41] Ladyzhensky OA, Salonnikov VA, Uraltseva NN. Linear and quasi-linear equations of parabolic type. Moscow: Science; 1967. 736 p.
[42] Budak BM, Samarskii AA, Tihonov AN. The collection of problems on mathematical physics. Moscow: Physics Mathematical Publishing; 1962. 687 p.
[43] Oben ZP, Ekland I. The applied nonlinear analysis. Moscow: World; 1988. 510 p.
[44] Grishin AM, Yakimov A. The method of solution of three-dimensional nonlinear boundary problems. Comput Technol 2003;8. Part 2 (Joint Release) (The Regional Bulletin of the East 3 2003;(19):176–85).
[45] Grishin AM, Pugacheva LV. Analytical solution of a problem of ignition of a wooden house wall as a result of forest fire. Bull Tomsk State Univ Math Mech 2010; 3(11):88–94.
[46] Perelygin LM, Ugolev BN. Wood-keeping. Moscow: Wood Industry; 1971. 286 p.
[47] Lykov AV. Heat and mass transfer. Handbook. Moscow: Energy; 1972. 560 p.
[48] Kudinov VA, Kudinov IV. Receipt and analysis of exact analytical solutions hyperbolic heat conduction equation for a flat wall. Teplophisics High Temp 2012;50(1):118–25.
[49] Baumeister K, Hamill T. Hyperbolic heat equation solution of the semi-infinite body. Heat Transm 1969;4:112–9.
[50] Jou D, Casas-Vazquez J, Lebon G. Extended irreversible thermodynamics. Berlin/Heidelberg: Springer-Verlag; 2001. 462 p.

[51] Podstrigach YS, Kolyano YM. Generalized thermomechanics. Kiev: Scientific Thought; 1976. 312 p.
[52] Kartashov EM. Integral relations for analytical solutions transport of hyperbolic models. News RAS Energ 2011;6:140–50.
[53] Koshkin NI, Shirkevich MG. Reference of elementary physics. Moscow: Science. Home Edition Physical and Mathematical Literature; 1988. 254 p.
[54] Grishin AM, Yakimov AS. Iteration-interpolation method solutions for three-dimensional wave equation. Comput Technol 2007;12(1):22–34.
[55] Yakimov AS. On a method of solving the conjugate heat transfer problem. Part 1. J Eng Phys 2013;86(3):453–63.
[56] Yakimov AS. On a method of solving the conjugate heat transfer problem. Part 2. J Eng Phys 2013;86(3):464–74.

# INDEX

Note: Page numbers followed by *t* indicate tables.